In Memoriam

Skyhorse Publishing books may be purchased in bulk at special discounts for sales promotion, corporate gifts, fund-raising, or educational purposes. Special editions can also be created to specifications. For details, contact the Special Sales Department, Skyhorse Publishing, 307 West 36th Street, 11th Floor, New York, NY 10018 or info@skyhorsepublishing.com.

Skyhorse® and Skyhorse Publishing® are registered trademarks of Skyhorse Publishing, Inc.®, a Delaware corporation.

Visit our website at www.skyhorsepublishing.com.

10 9 8 7 6 5 4 3 2

Library of Congress Cataloging-in-Publication Data is available on file.

Hardcover ISBN: 978-1-64821-073-0
Ebook ISBN: 978-1-5107-7643-2

Cover design by Brian Peterson
Book design by Geoff Towle

Printed in the United States of America

"Cause Unknown"

The Epidemic of Sudden Deaths In 2021 and 2022 and 2023

by Edward Dowd

Skyhorse Publishing

The measure of intelligence is the ability to change.

– Albert Einstein

STUDENT: *Dr. Einstein, aren't these the same questions as last year's final exam?*

DR. EINSTEIN: *Yes. But this year the answers are different.*

HOW TO USE THIS BOOK

Consider having a marker on hand as you read. Put a big **X** through anything you believe is not true. Next to that **X**, write what _is_ true. Without that second step, any of us might instantly discount or disregard information that challenges something we've assumed – often just because the new information is something we haven't heard before.

You might be inspired to seek out a fact-checking site to test or refute some information in this book that you find hard to believe. Fair enough. But before you automatically rely upon fact checkers, use the QR code to confirm that the Reuters CEO who started Reuters Fact Check was simultaneously on the Board of Pfizer – and still is – while simultaneously serving as Chairman of the Reuters Foundation that oversees their fact checking operation today. Conflicted and circular relationships like this are common when it comes to pharmaceutical marketing.

There's good reason Pfizer would want a board member with influence over Reuters Fact Check. Just like there's good reason they have a board member who was recently head of the FDA. And there's good reason for each of the people they added to their board in 2020: the outgoing CEO of the Gates Foundation (global public health relationships), and the current CEO of CocaCola (global distribution relationships). It's natural that Pfizer would do all it can to promote and protect the highest grossing consumer product in world history.

Also confirm that Pharma companies sponsor the overwhelming majority of TV news programs.

Also confirm that 75% of the funding for the FDA's drug division comes directly from Pharma

Also confirm that the government-funded British Broadcasting Corporation created Trusted News Initiative (TNI), the very active coalition of Reuters, The Washington Post, the Canadian and Australian Broadcasting Corporations, the European Broadcasting Union, Associated Press, Microsoft, Google, YouTube, Facebook and Twitter, all "_pledged to work together to **tackle harmful misinformation about Covid-19 vaccines**_."

TNI member companies work together to censor what they label as harmful information, while promoting and elevating information from public health institutions and Big Pharma. TNI members Google and Facebook provide funding to FactCheck.Org, earmarked to support COVID-19 coverage and "debunk viral deceptions." FactCheck.Org also receives funding from the Robert Wood Johnson Foundation (Johnson & Johnson family), the president of which served as acting-Director of the CDC, and was also Medical Editor at ABC News. These conflicted and circular relationships are at the center of creating and influencing public perception of pharmaceutical products.

That's why QR codes throughout these pages take you to original source material where you can confirm each report is real and each statistic well-sourced. Only when provided with more than one view can any of us decide for ourselves what's promotion, what's propaganda, what's marketing, and what is truth.

If we are not able to ask skeptical questions, to interrogate those who tell us that something is true, to be skeptical of those in authority, then we're up for grabs for the next charlatan – political or religious– who comes ambling along.

It wasn't enough, Jefferson said, to enshrine some rights in a constitution or a bill of rights. The people had to be educated, and they had to practice their skepticism... otherwise, we don't run the government, the government runs us.

– Carl Sagan

Those in authority are always right.

Their decision-making processes are always best.

They are never conflicted or corrupted.

Their views and opinions should always be accepted and followed.

Pharmaceutical companies have always prioritized health over profits.

Their safety trials are always sound and well-intentioned.

Their analysis of their own products should always be accepted.

– Some People

TABLE OF CONTENTS

FOREWORD

by Robert F. Kennedy, Jr.

Among the world's towering financial titans is BlackRock, a company with a bigger economy than every country on Earth except the U.S. and China. They manage $10 trillion in assets. In 2002, BlackRock recruited the brilliant Wall Street careerist Edward Dowd, and soon promoted him to serve as Managing Director. Turns out BlackRock made a very good bet on Ed Dowd: The Growth Fund he managed started at $2 billion; by the time he left BlackRock it was $14 billion.

His work with BlackRock required a keen ability to understand markets, pick stocks, analyze statistics, and identify trends.

In 2021, Dowd found himself withdrawing from Wall Street to study an entirely new kind of trend: the expanding and tragic epidemic of sudden deaths among healthy young people.

Every country dutifully maintains statistics on what's called All-Cause Mortality – deaths from any cause whatsoever. Whether accidents, disease, suicide, homicide, natural disaster or unexplained deaths, there is a long-established and fairly consistent baseline of All-Cause Mortality, year over year. Anything above that baseline is considered Excess Death. In 2021, it was Ed Dowd –not the public health officials that citizens rely upon– but Dowd who brought international attention to the fact that healthy working-age Americans were dying, and dying suddenly, at an alarming rate not seen before. These excess deaths were not anticipated by insurance actuaries, and weren't attributed to COVID.

Dowd framed the issue in a way I can't forget:

> "From February 2021 to March 2022, millennials experienced the equivalent of a Vietnam war, with more than 60,000 excess deaths. The Vietnam war took 12 years to kill the same number of healthy young people we've just seen die in 12 months."

One after another, reports from life insurance companies confirmed what Dowd was discovering, and in early 2022, he convened a group of insurance industry executives to explore it further. Later, he recruited expert analysts from around the world, and drawing on data from various official sources in many countries, he and his growing team committed to study the topic from every available vantage point.

In this unusual book, Ed Dowd proves an undeniable and urgent reality, laid out with facts that can be confirmed by any reader, point by point, page by page. He has helped us all understand something that many powerful people want to deny – and would get away with denying were it not for his skills and integrity.

Anyone who appreciates truth and accuracy owes Ed Dowd their gratitude. He certainly has mine.

RFK, Jr.

INTRODUCTION

In early 2021, I started hearing alarming and unusual stories of young, exceptionally fit athletes dropping dead on the field of play. If those who are the healthiest among us are suddenly dying, what would that mean for the rest of the population? In other words, what if healthy young athletes are the canary in the coalmine? To my trained eye, what others might consider sad anecdotes became something more: A trend change was underway.

Success during my career came during those times I was able to identify trend changes before they were apparent to my peers and the financial press.

Using similar skills, I can look at today's health and death statistics and see that something isn't right. To be more specific, something about All-Cause Mortality and the excess death rate isn't right. Determining All-Cause Mortality is not controversial. It's just math. And it's the kind of math that insurance actuaries obsess upon, for understandable professional reasons. They scrutinize total deaths by age group, by ethnicity, by region, by year, even day to day. They studiously compare month-to-month and year-to-year stats, searching for trends. Usually, the rate of deaths is fairly consistent. For example, in 1933 about 1.4 million Americans died from all causes. In 1955, about 1.5 million Americans died from all causes, and about the same number again in 1956.

By 2017, around 2.8 million Americans died. 2018 was about the same again. 2019, about the same again. Not surprisingly, 2020 saw a spike, smaller than you might imagine, some of which could be attributed to COVID and to initial treatment strategies that were not effective.

But then, in 2021, the stats people expected went off the rails. The CEO of the OneAmerica insurance company publicly disclosed that during the third and fourth quarters of 2021, death in people of working age (18-64) was 40% higher than it was before the pandemic. Significantly, the majority of the deaths were not attributed to COVID.

A 40% increase in deaths is literally earth-shaking, and not only for the devastated families and communities that directly experience the deaths. Even a 10% increase in excess deaths would have been a 1-in-200-year event. But this was 40%.

And therein lies a story – a story that starts with obvious questions:

What has caused this historic spike in deaths among younger people?

What has caused the shift from old people, who are expected to die, to younger people, who are expected to keep living?

It isn't COVID, of course, because we know that COVID is not a significant cause of death in young people.

Various stakeholders will opine about what could be causing this epidemic of unexpected sudden deaths. Though I'll share my best conclusions, I aim most of all to help you reach your own conclusions. In the coming pages, I won't ever ask you to rely upon me for anything; all the facts I share will have citations you can quickly confirm. I won't be expressing mere opinions or making arguments. The facts just are, and the math just is.

Though excess and All-Cause Mortality can be expressed in numbers, I begin this book with the actual human reality behind those numbers. As you see some of the actual people who are represented by the dry term Excess Mortality, it's difficult to take on board the unexpected sudden deaths of young athletes, known to be the healthiest among us. Similarly, when lots of healthy teenagers and young adults die in their sleep without obvious reason, collapse and die on a family outing, or fall down dead while playing sports, that all by itself raises an immediate public health concern. Or at least it used to.

As you turn the next few pages, ask yourself if you recall seeing these kinds of things occurring during your own life – in junior high? In high school? In college? How many times in your life did you hear of a performer dropping dead on stage in mid-performance? Your own life experience and intuition will tell you that what you're about to see is not normal.

Or at least it wasn't normal before 2021.

Ed Dowd

HEALTHY YOUNG ATHLETES...

NEWS REGION SPORT SHOW PLAY PODCAST PUZZ

Wouter (14) collapses on hockey field and dies, school and club in mourning

Sebastian Quekel 08-12-21, 13:49   Last update: 08-12-21, 15:58

CONFIRM THIS STORY IS REAL

DAILY NEWS

College basketball star Derek Gray dead at 20 after collapsing at campus basketball camp

By Jami Ganz · New York · Jul 30, 2022 at 4:57 pm

CONFIRM THIS STORY IS REAL

HLN   NEWS SPORT SHOWBIZ NINA NEAR VIDEO PUZZLE PODCA

Football player Niels De Wolf (27) died after being struck by heart failure after a game on Sunday

Kristof Pieters 07-10-21, 08:28   Last update: 07-10-21, 09:12

CONFIRM THIS STORY IS REAL

fanpage.it

JANUARY 26, 2022   14:45

Treviso, Carlo Alberto died: the 12-year-old athlete suffering from cardiac arrest during a race

CONFIRM THIS STORY IS REAL

Olé

Get into   SUB

The chilling death of a 17-year-old player during a match

CONFIRM THIS STORY IS REAL

SUDINFO •

Provincial football in mourning: Kévin, 26, collapses in the middle of training

Posted on Wednesday, August 17, 2022 at 12:13 p.m. By VC

CONFIRM THIS STORY IS REAL

SportoweFakty

Tragedy during training. The 22-year-old is dead

Mateusz Kozanecki
June 16, 2022, 12:56

CONFIRM THIS STORY IS REAL

22-Year-Old Man Dies While Competing in a Marathon

By Joelle Goldstein | Published on October 4, 2021 01:50 PM

CONFIRM THIS STORY IS REAL

DH SPORTS

A Fémina Visé player dies suddenly at the age of 19

Whitnée Abriska died after going into cardiac arrest.

Published on 2021-07-26 at 5:00 p.m. - Updated on 2021-07-27 at 08:21 a.m.

CONFIRM THIS STORY IS REAL

 TOTALWATERPOLO

Romanian water polo player, aged 23, died during game

March 26, 2022

Andrei Drăghici

CONFIRM THIS STORY IS REAL

l'avenir

a Michaël Englebert, 37, dies following a heart attack after his match with Ortho

Published on 2021-10-25 at 22:01

CONFIRM THIS STORY IS REAL

Croatian footballer Marin Cacic tragically dies aged 23 following collapse in training

By Josh O'Brien, Sports Writer
07:30, 24 Dec 2021 | UPDATED 13:32, 24 Dec 2021

CONFIRM THIS STORY IS REAL

15-year-old soccer player dies during tournament

By Rafael Oliveira and Adriel Morais, g1 Goiás and TV Anhanguera

01/12/2022 16:37 · Updated 7 months ago

CONFIRM THIS STORY IS REAL

BERGENFIELD

DAILY VOICE — DUMONT · NEW MILFORD

Dumont Boy, 14, Collapses, Dies Playing Basketball

Jerry DeMarco   11/21/2021 11:30 a.m.

CONFIRM THIS STORY IS REAL

W Wales Online + Follow   View Profile

Young rugby player who 'gave his heart' to the game dies suddenly aged 17

Katie-Ann Gupwell - 2 Nov 2021

Rugby Community Mourns Sudden Death Of Player

TRIBUTES HAVE BEEN LEFT AT

CONFIRM THIS STORY IS REAL

'Gentle giant' dad, 29, collapses and dies after rugby match leaving fans shocked

By Jim Hardy & Matthew Dresch, News Reporter

02:07, 8 Sep 2021

Mirror

CONFIRM THIS STORY IS REAL

NEW YORK POST LOG

Canadian junior hockey captain dies during tournament game

By Alec Gearty   September 1, 2022 | 3:09am | Upda

Canadian junior hockey captain dies during game

519 SPORTS ONLINE

CONFIRM THIS STORY IS REAL

DAILY NEWS HUNGARY
Politics   Business   Society   Sport   Culture   Special Hung

Péter Licskay - 23/06/2021 · Sport

Tragedy! – Young Hungarian footballer dies on the field

ARR ANDRÁSHIDA S

CONFIRM THIS STORY IS REAL

Going back to Ancient Greece, young athletes have been championed as the flower of the human race, more celebrated than philosophers, scientists, and artists. When we gather together to watch them perform at the highest level, we share a communal emotion akin to a religious experience. A young athlete's death is therefore perceived as the most difficult to comprehend and come to terms with.

— John Leake

Young father and footballer dies of heart attack during training

Rachel Moore · 17:23, Aug 23 2021

CONFIRM THIS STORY IS REAL

World Today News

At 22, James Théodore, Rugby player, dies in training

News · March 1, 2022 · No Comments

CONFIRM THIS STORY IS REAL

HOME | NEWS | ENTERTAINMENT | LIFESTYLE | MONEY | HEALTH

TRIBUTES PAID Mukhaled Al-Raqadi dies aged 29 after collapsing in warm-up

Tristan Barclay
5:48 ET, Dec 23 2021 | Updated: 8:17 ET, Dec 23 2021

CONFIRM THIS STORY IS REAL

EuroWeekly NEWS
NEWS AND VIEWS SINCE 1998

Second Oman footballer dies following a heart attack at 30

By Matthew Roscoe · 27 January 2022 · 15:00

CONFIRM THIS STORY IS REAL

Stillwater County NEWS

Park City football player dies after collapsing on field

— Marlo Pronovost, SCN Editor   Thursday, November 4, 2021

CONFIRM THIS STORY IS REAL

B B C   Home   News   Sport   Reel   ···   Q

NEWS

England | Regions | Nottingham

Boy, 13, dies after collapsing at Nottinghamshire football match

8 May, 2022

CONFIRM THIS STORY IS REAL

Josh Downie: Cricketer, 24, dies after heart attack at practice

10 May 2021

CONFIRM THIS STORY IS REAL

Algerian football player Sofiane Lokar dies of heart attack during match

25 Dec 2021   20 💬

CONFIRM THIS STORY IS REAL

Mirror
FA Youth Cup footballer dies after 'suffering cardiac arrest' during match

By Darren Wells, Digital Football Writer
16:25, 5 Sep 2021 | UPDATED 16:42, 5 Sep 2021

CONFIRM THIS STORY IS REAL

THE NEW INDIAN EXPRESS

Hockey player dies in the middle of a game; heart attack suspected

Published: 25th December 2021 05:27 PM | Last Updated: 28th December 2021 01:34 PM

CONFIRM THIS STORY IS REAL

SPORT
Elite runner dead – suffered cardiac arrest during race in Kalmar

Erik Karlsson turned 23: "You are always with us"

PUBLISHED: 04 JANUARY 2022 AT 21.17

CONFIRM THIS STORY IS REAL

Crime

Pa. boy, 12, collapses, dies at middle school basketball practice

Updated: Nov. 01, 2021, 7:05 a.m. | Published: Oct. 28, 2021, 10:37 a.m.

CONFIRM THIS STORY IS REAL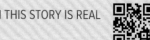

la Repubblica

MARCH 13, 2022   1 MINUTE READ   UPDATED MARCH 13, 2022 AT 11:46

Bolzano, hockey player Andreas Palla dies during a match

CONFIRM THIS STORY IS REAL

 EuroWeekly NEWS

Spanish footballer dies aged 41 following a half-time heart attack

By Matthew Roscoe · 16 February 2022 · 10:27

CONFIRM THIS STORY IS REAL

Evening Standard

Cyclist dies after suffering cardiac arrest during RideLondon charity bike ride

23 June 2022

CONFIRM THIS STORY IS REAL

EuroWeekly NEWS
NEWS AND VIEWS SINCE 1998

Shock as goalkeeper dies following mid-game heart attack

By Matthew Roscoe · 08 August 2022 · 16:39

CONFIRM THIS STORY IS REAL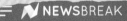

wbur
 DONATE

LOCAL COVERAGE

Home // Local Coverage

Newton teen dies weeks after collapsing during basketball game

March 01, 2022   By The Associated Press   🐦 f ✉

A Massachusetts teenager has died after going into cardiac arrest and collapsing while playing in a high school basketball game in early February, his family said Monday.

CONFIRM THIS STORY IS REAL

≡ ◈ NEWSBREAK   📱

PA High School Senior Athlete Dies Of 'Sudden Cardiac Incident,' Report Says

Daily Voice   2021-11-01

CONFIRM THIS STORY IS REAL

IndyStar.

North | Sports | Indy 500 | Things To Do | Opinion | Obituaries | E-Edition | Legals

Hanover basketball player, Danville grad, dies unexpectedly

Matthew VanTryon
Indianapolis Star

Published 3:54 p.m. ET Oct. 19, 2021 | Updated 4:35 p.m. ET Oct. 20, 2021

CONFIRM THIS STORY IS REAL

gl — SANTOS AND REGION

31-year-old athlete dies after heart attack during race in Santos

12/06/2021 1:17 pm · Updated8 months ago

CONFIRM THIS STORY IS REAL

De Telegraaf
MENU | Subscribe

Dutchman (19) suddenly died during a walk

Aug 07, 2022 in INTERIOR

CONFIRM THIS STORY IS REAL

24 ≡

Tragedy: The player collapsed on the field and died!

Written by Zdravko Barišić, Monday, April 12, 2021. at 2:35

CONFIRM THIS STORY IS REAL

T13 — en vivo (•)

Renato Bastías: the triathlete who died of a heart attack at the Pucón Ironman

MONDAY, JANUARY 10, 2022 6:15 P.M.

CONFIRM THIS STORY IS REAL

≡ The Mercury News 👤∨ (

Clayton middle school student dies following youth football practice

PUBLISHED: August 9, 2022 at 11:01 a.m. | UPDATED: August 10, 2022 at 9:31 a.m.

CONFIRM THIS STORY IS REAL

Young soccer player dies of a heart attack after a match

BYABC COLOR
NOVEMBER 07, 2021, - 22:25

CONFIRM THIS STORY IS REAL

Belgian amateur cyclist dies of heart attack when training in Mallorca

MATT HANSEN MARCH 24, 2022

CONFIRM THIS STORY IS REAL

CARRERAS PORMONTAÑA

Josep Maria Pijuan dies in La Llanera Trail 2022

News ⏱ Monday, January 24, 2022 - 7:22 pm

CONFIRM THIS STORY IS REAL

Noticia y PUNTO

Young athlete dies while playing basketball with her friends

· February 7, 2022 💬 0 🔥 2,228 🏳 1 minute read

CONFIRM THIS STORY IS REAL

NEW YORK POST

Sports Sports Betting SPORTS+ NFL MLB NBA NHL NCAA

Undefeated boxer Musa Askan Yamak dies of heart attack during fight

By Joshua Lynch   May 18, 2022 | 6:23am | Upda

CONFIRM THIS STORY IS REAL

la Repubblica

Viterbo, thirty years old, dies while playing soccer.

NOVEMBER 12, 2021   ⏱ 1 MINUTE READ

CONFIRM THIS STORY IS REAL

Questions surround Lawrence North student's death after collapse at practice

by: Steve Brown — Chief Investigator
Posted: Jun 8, 2022 / 03:55 PM EDT
Updated: Jun 8, 2022 / 06:03 PM EDT

CONFIRM THIS STORY IS REAL

Kennedy High Football Player Emmanuel Antwi Dies After Collapsing During Game

MARCH 22, 2021 5:55 AM | CBS SACRAMENTO

Kennedy High Athlete Dies After Collapsing During Game

CONFIRM THIS STORY IS REAL

Student-athlete Cameran Wheatley collapses during basketball game, dies at hospital

By Elizabeth Matthews | Published February 9, 2022 | Midlothian | FOX 32 Chicago

CONFIRM THIS STORY IS REAL

NEWS

Texas High School Basketball Player DeVonte Mumphrey Dies During Game

BY GERRARD KAONGA ON 2/9/22 AT 8:55 AM EST

NEWS   BASKETBALL   TEXAS

A student at Alto High School, Texas, has died after collapsing during a basketball game.

CONFIRM THIS STORY IS REAL

Brit footballer, 35, dies 'after suffering heart attack on the pitch' while in Dubai

By James Rodger, Regional Content Editor & Tim Hanlon, News Reporter
03:03, 29 Jan 2022 | UPDATED 07:50, 29 Jan 2022

CONFIRM THIS STORY IS REAL

Ypsilanti student dies after medical emergency during basketball tryouts

Published: November 16, 2021 at 3:06 PM

CONFIRM THIS STORY IS REAL

You can hold out as long as you want but you won't have much freedom. I'm over it, I did it. Does it make me a sheep? No.

— Jake Kazmarek, September 28, 2021
Died Four Days After Second Vaccine

In Loving Memory
Jake R. Kazmarek
January 24, 1993 - October 2, 2021

CONFIRM THIS STORY IS REAL

Haitem Jabeur Fathallah dies after an illness in the match

Sport · October 17, 2021 · No Comments

CONFIRM THIS STORY IS REAL

alagoas 24 horas · Sunday 08/28/2022 · see all

Young man has a heart attack on the soccer field and arrives at the hospital lifeless

01/30/2022 08:17 | leave a comment

Adriano's death is the second in the city, under identical circumstances, in just over 24 hours. On the morning of Friday, the 28th, the ex-player of the Jaciobá and Internacional teams, Ivo Santos, 39, had his death confirmed shortly after suffering a sudden illness while playing futsal at a local sports court.

CONFIRM THIS STORY IS REAL

Footballer dies aged 25 after collapsing in training preparing for new campaign

By Sam Fletcher
11:27, 4 JAN 2022 | UPDATED 11:29, 4 JAN 2022

CONFIRM THIS STORY IS REAL

STAR'S TRAGEDY Ex-Parma footballer Giuseppe Perrino dies aged 29 at brother's memorial match after collapsing while playing

Marc Mayo

12:30, 3 Jun 2021 | Updated: 12:49, 3 Jun 2021

CONFIRM THIS STORY IS REAL

Daily Mail .com

Tragedy as ex-Denver Broncos offensive lineman dies at just 35 after going into cardiac arrest on a run

PUBLISHED: 10:49 EDT, 22 July 2022 | UPDATED: 11:12 EDT, 25 July 2022

CONFIRM THIS STORY IS REAL

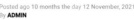

REGIONAL

Soccer player suffered a heart attack during practice

Posted ago 10 months the day 12 November, 2021
By ADMIN

CONFIRM THIS STORY IS REAL

Jens de Smet dies after collapsing on pitch during football game

DECEMBER 5, 2021 08:14
by DZEVAD MESIC | VIEW 4771

CONFIRM THIS STORY IS REAL

People

SUBSCRIBE

18-Year-Old College Football Player Dies After Collapsing During Practice

By **Greta Bjornson** | Published on August 10, 2021 03:08 PM

CONFIRM THIS STORY IS REAL

B B C Home News Sport Reel Worklife ···

Cwmllynfell rugby player Alex Evans dies during match

🕒 22 August 2021

CONFIRM THIS STORY IS REAL

EL NACIONAL

Athlete Alexaida Guédez died during a race in Naguanagua

By The National · August 24, 2021

CONFIRM THIS STORY IS REAL

DAILY MAVERICK
SPORT

Hockey player collapses, dies on field

CONFIRM THIS STORY IS REAL

4 NEW YORK   SHARE

NY Student With No Prior Health Issues Collapses, Dies at Basketball Tryout

By Greg Cergol • Published November 18, 2021 • Updated on November 19, 2021 at 3:28 am

CONFIRM THIS STORY IS REAL

Never seen anything like it. I played for 17 years. I don't think I saw one person in 17 years have to come off the football pitch with breathing difficulties, clutching their heart, heart problems. But, in the last year, it's just been unbelievable how many people, not just footballers, sports people in general – tennis players, cricketers, basketball players, just how many are just keeling over. And at some point, surely you have to say, 'This isn't right, this needs to be investigated.'

— Matt Le Tissier, Global Star Footballer

Football game canceled following death of player who collapsed at practice

Published: Aug. 25, 2021 at 7:19 AM PDT | Updated: Aug. 25, 2021 at 5:35 PM PDT

STUDENT DIES AFTER COLLAPSING AT FOOTBALL PRACTICE
UPCOMING DUTCH FORK HIGH SCHOOL GAME CANCELLED

CONFIRM THIS STORY IS REAL

Sport > Premier League

RIP ALEKO Alexandros Lampis dead at 21: Greek footballer dies after cardiac arrest

12:43 ET, Feb 2 2022 | Updated: 13:22 ET, Feb 2 2022

CONFIRM THIS STORY IS REAL

EL ORBE

Soccer player dies of a heart attack

February 9, 2022   835

CONFIRM THIS STORY IS REAL

KAUMUDI ONLINE

18-year-old boy dies after collapsing during football game

Wednesday 15 September, 2021 | 12:55 AM

CONFIRM THIS STORY IS REAL

Irish Mirror

Tributes pour in for young footballer with 'a heart of gold' after sudden death

14:28, 22 AUG 2022 | UPDATED 08:55, 24 AUG 2022

CONFIRM THIS STORY IS REAL

DAILY VOICE BERKS

Exeter Township High School Senior Dies Suddenly

Nicole Acosta   01/06/2022 10:58 a.m.

CONFIRM THIS STORY IS REAL

NewportRI.com | The Newport Daily News

'International presence': Ex-America's Cup sailor dies while competing in Newport regatta

Published 12:35 p.m. ET June 8, 2022

CONFIRM THIS STORY IS REAL

 WFTS TAMPA BAY

Citrus High School football player dies after collapsing at practice

Posted at 9:29 PM, Sep 28, 2021

Photo: Men of Vision

CONFIRM THIS STORY IS REAL

AGP

ATHLETICS

16-year-old athlete dies on Sullivan track

Published 11 months ago on 12 October 2021
By **AGP News**

CONFIRM THIS STORY IS REAL

Le Petit Bleu

Subscr

Menu

Death of Jérôme Garens after a rugby match: the world of rugby in shock

SUBSCRIBERS

CONFIRM THIS STORY IS REAL

Telegraf

20/09/21 | 09:31 > 20:11

Young football player's pulse returned briefly, but he couldn't be saved and passed away in seconds

CONFIRM THIS STORY IS REAL

en English

Boy, 14, dies 5 days after cardiac arrest on soccer pitch

07 September 2022   18:10   NEWS

OSPEDALE CIVILE

CONFIRM THIS STORY IS REAL

oBristolWorld

News

Student collapsed and died in PureGym after work-out

24th Oct 2022, 10:58pm

CONFIRM THIS STORY IS REAL

india.com

East Bengal FC Bound Debojyoti Ghosh Dies Due to Heart Attack

Published: March 20, 2022 1:28 PM IST

CONFIRM THIS STORY IS REAL

CONTRA FATOS !

Heart Attack Kills Muay Thai Champion At 32

Per Counterfacts
Published in 03/01/2022

CONFIRM THIS STORY IS REAL

el Periódico   kiosk

A 27-year-old athlete from Tauste dies after suffering a cardiac arrest

Saragossa | 17 11 21 | 19:05 | Updated at 19:28

CONFIRM THIS STORY IS REAL

REUTERS   World ⌄ Business ⌄ Legal ⌄ Markets ⌄ More ⌄

June 14, 2021
10:15 AM PDT
Last Updated a year ago

Sports

Indonesian doubles star Kido dies of heart attack at 36

Reuters

1 minute read

CONFIRM THIS STORY IS REAL

Sahil Online
REFLECTION OF THE TRUTH

Kabaddi player Manoj Naik dies of heart attack

Source: DHNS | Published on 22nd September 2021, 12:30 PM

CONFIRM THIS STORY IS REAL

LA NOTIZIA

Maria Sofia Paparo, the swimming champion struck down by a heart attack

Posted on April 12, 2022 by Fabrizio Capecelatro

CONFIRM THIS STORY IS REAL

Patch   Sign

Football Player Dies After Collapsing On College Campus

Posted Fri, Aug 12, 2022 at 1:58 pm CTUpdated Fri, Aug 12, 2022 at 2:14 pm CT

CONFIRM THIS STORY IS REAL

Bristol World

Devastated mother says 'her beautiful boy' who suddenly died from heart attack should still be alive

Friday, 29th July 2022, 10:38 pm

CONFIRM THIS STORY IS REAL

JOGADA 10

26-year-old Brazilian player dies of cardiac arrest

POSTED 6 MONTHS AGO

CONFIRM THIS STORY IS REAL

ALCALAHOY

A young footballer dies suddenly

By Alcala Today - January 23, 2022

CONFIRM THIS STORY IS REAL

Mirror

Cavan community pays tribute to 'hugely talented' young Irishman after sudden death

By Michelle Cullen michelle.cullen@reachplc.com
10:19, 25 JAN 2022   UPDATED 17:51, 25 JAN 2022

CONFIRM THIS STORY IS REAL

> *Get your double vax and get on with it and learn to live with it.*
>
> — Shane Warne, 2021

NEW YORK POST

Australian cricket legend Shane Warne dies in his sleep

By Emily Crane

March 4, 2022 | 10:07am | Updated

Shane Warne -- survived by his three children -- died of a suspected heart attack.

CONFIRM THIS STORY IS REAL

37-year-old cycling champ dies days after winning national race

Aug. 24, 2022, 6:29 PM PDT

By Phil Helsel

Scottish cyclist Rab Wardell, 37, died unexpectedly Tuesday after he suffered cardiac arrest, his partner said.

Wardell went into cardiac arrest Tuesday morning while the couple were lying in bed, Katie Archibald, an Olympic medalist in track cycling, said on social media Wednesday.

"I tried and tried, and the paramedics arrived within minutes, but his heart stopped and they couldn't bring him back," she wrote. "Mine stopped with it. I love him so much and need him here with me. I need him here so badly, but he's gone. I can't describe this pain."

Scottish Cycling confirmed the death.

CONFIRM THIS STORY IS REAL

:NEWS

Andrea Musiu, a young promise, dies at the age of 20

Posted on:24-07-2022 16:57

CONFIRM THIS STORY IS REAL

Pioneer
North Wales

Tragedy as Colwyn Bay Under 16s goalkeeper dies suddenly

15th March

CONFIRM THIS STORY IS REAL

THE IRISH Sun

TRAGIC DEATH Tributes to 'bright and funny' Irish fitness coach Scott Murray MacDonald who died suddenly at his home

Fionnuala Walsh

12:35, 3 Mar 2022 | Updated: 12:35, 3 Mar 2022

CONFIRM THIS STORY IS REAL

EuroWeekly NEWS
NEWS AND VIEWS SINCE 1998

Tributes pour in after sudden death of young athlete in Ireland

By Joshua Manning • 21 June 2022 • 13:44

CONFIRM THIS STORY IS REAL

FOX NEWS

Login | Watch TV

Published August 18, 2022 10:26am EDT

FIU linebacker Luke Knox dead at 22

CONFIRM THIS STORY IS REAL

BLACK ENTERPRISE THE #1 Black Digital Media Brand

ALABAMA STATE SOPHOMORE LINEBACKER AWYSUM HARRIS FOUND DEAD IN DORM ROOM

Cedric 'BIG CED' Thornton ⏲ July 5, 2022 👁 60754

REMEMBERING
AWYSUM HARRIS
2001 - 2022

CONFIRM THIS STORY IS REAL

TMZ SPORTS

CANADIAN COLLEGE FOOTBALL PLAYER
FRANCIS PERRON DEAD AT 25

9/20/2021 9:06 AM PT

CONFIRM THIS STORY IS REAL

batln

Article published on 05.10.2021 at 06:00

Mourning for Benjamin Taft: An unbelievable, far too early farewell

CONFIRM THIS STORY IS REAL

the POST

YOUNG NAT'L BASEBALL PLAYER JEROME YENSON DIES AT 24

JANELLE MENESES / 25 SEPTEMBER 2021

CONFIRM THIS STORY IS REAL

The U.S. Sun

HOME | NEWS | ENTERTAINMENT | LIFESTYLE | MONEY | HEALTH | SPORT | >

Premier League > Soccer

RIP MARCO Italian-born teen footballer Marco Tampwo dead aged 19 after suspected heart attack

Steve Goodman
9:00 ET, Aug 17 2021 | Updated: 9:18 ET, Aug 17 2021

CONFIRM THIS STORY IS REAL

Comrades Marathon returning with 15 000 fully-vaccinated runners

17 Feb 2022

msn

Comrades Marathon | Concern over 'abnormal' deaths

JOHANNESBURG - At least 74 athletes had to be transferred to hospitals after struggling during the Comrades Marathon.

Two runners namely Mzamo Mthembu and Phakamile Ntshiza tragically died.

Two other athletes were in the Intensive Care.

Note: The temperature was less than 70 degrees

Running   SUBSCRIBE

Runner collapses and dies at Comrades Marathon

MARLEY DICKINSON   AUGUST 29, 2022

Times LIVE

Second Comrades Marathon runner loses his life

30 August 2022 - 10:35

CONFIRM THIS STORY IS REAL   CONFIRM THIS STORY IS REAL

News   Weather   Near Me   VERIFY

LOCAL NEWS

GBI: Southwest player died of heart problem

15-year-old Joshua Ivory Jr. died shortly after collapsing at football practice in late July.

Author: 13WMAZ Staff
Published: 1:02 PM EDT November 4, 2021
Updated: 1:42 PM EDT November 4, 2021

CONFIRM THIS STORY IS REAL

INTERNATIONAL ICE
HOCKEY FEDERATION

Sadecky suffers fatal attack

by Andrew Podnieks | 05 NOV 2021

CONFIRM THIS STORY IS REAL

Sun
SOUTH COAST

Amanzimtoti canoeist dies after suspected heart attack

October 11, 2021   Bianca Lalbahadur   1 minute read

CONFIRM THIS STORY IS REAL

CONNECT WITH US

Third Division's Rabat & Anwar goalkeeper dies of cardiac arrest

by Ahmad Gamal Ali   December 23, 2021

CONFIRM THIS STORY IS REAL

srednja.hr

Filip Turk, a 22-year-old cadet football player of the local football club, died suddenly.

CONFIRM THIS STORY IS REAL

SPORT NEWS AFRICA

9 months | At a glance , Football   December 27, 2021

Italy: Adrien Sandjo dies after a cardiac arrest

CONFIRM THIS STORY IS REAL

Sopore MBBS student dies of heart attack

January 13, 2022

CONFIRM THIS STORY IS REAL

16-year-old Mississippi football player dies

Published: Nov. 8, 2021 at 8:19 PM PST

CONFIRM THIS STORY IS REAL

Cincinnati.com | The Enquirer

HIGH-SCHOOL-SPORTS

Northwest, North Carolina A&T volleyball's Fatimah Shabazz dies

Scott Springer
Cincinnati Enquirer

Published 4:02 p.m. ET Nov. 30, 2021 | Updated 9:46 p.m. ET Dec. 1, 2021

CONFIRM THIS STORY IS REAL

LOCAL 12

Wyoming High School graduate, Kentucky Wesleyan freshman dies unexpectedly

by WKRC
Sunday, January 2nd 2022

CONFIRM THIS STORY IS REAL

THE HINDU

SPORT · CRICKET

CRICKET

Young Saurashtra cricketer Avi Barot dies after suffering cardiac arrest

PTI

RAJKOT OCTOBER 16, 2021 14:16 IST
UPDATED: OCTOBER 16, 2021 14:16 IST

CONFIRM THIS STORY IS REAL

HA Hellweger Anzeiger

Mourning for Lukas Bommer (25): The goalkeeper who always laughed is dead

untains /10/12/2021, last updated15:49, 10/11/2021/ Reading time: 2 minutes

CONFIRM THIS STORY IS REAL

FOX5 ATLANTA
Norcross High School mourns death of football player
By FOX 5 Atlanta Digital Team | Published November 11, 2021 | Norcross | FOX 5 Atlanta

TENERIFE WEEKLY
Shock over the death of the young Canarian athlete Mateo Hernández
January 13, 2022   La Provincia   Reading Time: 2 mins read   💬 0

THE U.S. Sun
Sport
RIP David Jenkins dead at 31
Gary Stonehouse

2:30 ET, Oct 12 2021 | Updated: 3:04 ET, Oct 12 2021

WLR
Tipperary GAA in mourning following sudden passing of Dillon Quirke (24)
August 06, 2022 7:31 AM

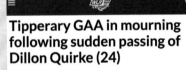

the japan times
J. LEAGUE
Bellmare midfielder Riuler dies at 23
KYODO   SHARE   Nov 24, 2021

nam! strzyżów naszemiasto
Krystian Kozek, 17-year-old player of Wisłok Strzyżów, is dead
October 30, 2021, 8:24

fanpage.it
DECEMBER 5, 2021   14:56
24-year-old runner Fabio Pedretti suddenly dies
Edited by Simona Buscaglia

aragondigital.es
Dani Gómez, 18-year-old Peñas Huesca basketball player, dies
23 AUGUST, 2022

WLWT5
87° WEATHER
Cincinnati high school volleyball player, cheerleader dies
WLWT5   Updated: 5:19 PM EDT Mar 14, 2022

WESTERN HILLS HIGH SCHOOL MOURNING

People
SUBSCRIBE
UMass Amherst Lacrosse Player Dies 'Unexpectedly' at 19
By Glenn Garner   Published on June 2, 2022 10:54 PM

nam! gliwice naszemiasto
Company catalogue   Contact to the editorial office   Mor
Michał Kapias is dead. He was only 22 years old
DEVELOPED: Szymon Bijak   February 13, 2022, 7:47

"Sikret" Swimming Club Gliwice / Facebook

Belfast Telegraph
Dominic Oscar: West Belfast boxer (19) who died suddenly had 'kind-hearted nature' and 'huge potential'
August 14 2022 08:25 PM

Daily **Mail**.com

Hockey player, 16, dies after suffering a series of strokes

By KEITH GRIFFITH FOR DAILYMAIL.COM
PUBLISHED: 02:04 EST, 28 December 2022 | UPDATED: 09:18 EST, 28 December 2022

3 ☰ News Watch Live   🔍

'Unimaginably heartbreaking': former JSU basketball player passes away

Published: Nov. 9, 2022 at 10:23 AM PST

CNN   Live TV ☰

UFC Hall of Famer Stephan Bonnar dies at 45

Updated 10:48 AM EST, Sun December 25, 2022

FOX NEWS   Login   Watch TV   ☰

Air Force football player Hunter Brown dead at 21   January 10, 2023

TMZ SPORTS

FORMER NFL GUARD DEAD AT 38
... Acute Heart Failure   1/2/2023 1:44 PM PT

🦚 NEWS

SPORTS

Victoria Lee, rising MMA star, dies at 18

Jan. 9, 2023, 4:32 AM PST / Source: TODAY

☰   People   SUBSCRIBE

Keith Farmer, Four-Time British Motorcycling Champion, Dead at 35

By Ingrid Vasquez | Published on November 10, 2022 06:09 PM

abc

High school student collapses, dies after suffering medical emergency during flag football

Tuesday, January 10th 2023, 8:16 AM CST

You've just seen a very small sampling of sudden deaths during 2021 / 2022

A sampling of many hundreds more appears in the Compendium on Page 145

FIRST, IS IT TRUE?

FIRST, IS IT TRUE?

Before 2021, none of us had ever seen anything like what you've just seen, and certainly never to this extent. If you were surprised there were so many confirmed news reports about healthy young athletes collapsing and dying in mid-play, please note that what you just saw is only a small sampling. Even all the tragedies confirmed in this book's Compendium are only a small sampling. And every one of the hundreds of people whose deaths are memorialized at the end of this book died younger than 46 years old.

At this point, I ask that only one thing be fully accepted as fact: That these deaths are real, are happening. And don't focus yet on why.

People and institutions hoping to suppress the information in this book will likely launch a criticism along the lines that looking at these deaths somehow disrespects the families. The exact opposite is true: ***Ignoring these deaths is the greatest disrespect we could ever show***. These young people and their families and communities deserve to have someone care about why they died – particularly when official agencies have closed their eyes (and perhaps their hearts) to such cares.

Every single health official in America knows the obvious: Healthy young people don't just stop living for no reason. The deaths in this book that happened at professional sports events are doubly confounding because those athletes died despite the immediate on-site presence of emergency medical responders who are trained and equipped to resuscitate people. But since 2021, so many could not be resuscitated; they died on the spot, arrived at the hospital already dead. That is both tragic and highly unusual.

Those seeking to discount the facts in this book might say there have always been cases of professional athletes dying from cardiac arrest during competition. And that's true. But has it ever been this often? Or so many victims this young?

The most cited study on the topic of sudden death in young athletes was done in 2006 at the Division of Pediatric Cardiology, University Hospital of Lausanne, Switzerland. Experts there looked at sudden cardiac death in athletes under 35-years old, between the years 1966 and 2004.

After a systematic review of the literature, the Lausanne Study determined there had been 1101 such cases – **over a period of 38 years**. That's an average of about 29 per year. These days, we'd be grateful to see a single month with only 29 such events. In fact, since June 2021, there hasn't been even one month with fewer than 29 of these sudden deaths; there were 90 reported in December 2021 alone, and about the same the month after that.

See Appendix One on Page 175 for more than 50 scholarly writings about sudden athlete deaths; none show anything approaching the number publicly reported since 2021.

If you believe athlete deaths have always happened at this rate and that nothing unusual is underway, conduct your own Google search for deaths like these in 2019, '18, '17, or any other past years. Try every combination of every search term you can think of (e.g., athlete sudden death, died midgame, young athlete collapsed and died, collapsed on field, suddenly died), and see if you can find even a fraction of what's occurred since 2021.

Healthy teenagers dying in their sleep – an event like that was beyond rare, which is why most of us lived our whole lives never hearing about such things. Until 2021.

Imagine that two healthy teenage boys are found dead in their beds a few days apart. In both cases, emergency responders knew immediately that resuscitation would be futile. Just days before they died, both had received the second dose of Pfizer's COVID vaccine. An autopsy in Connecticut, and another in Michigan confirmed the official cause of death for both boys: ***Vaccine-induced myocarditis*** (inflammation of the heart muscle, swelling of the heart).

When official agencies in two states conclude that COVID vaccines can kill healthy teenagers, wouldn't that seem to be important national news? But it was barely reported anywhere. Using the QR code, you can confirm right now that this suppressed story is true.

When a sudden death occurs in front of spectators, it's newsworthy, and will be reported at least in local media outlets. Similarly, when a healthy young famous person dies unexpectedly, it's likely to be reported by the news media. In the past, such news stories about a young person would keenly focus on answering the glaring question: *Why did this person die?* But the current protocol in national news is to ignore such details outright. In the rare instances that these deaths are even reported beyond local news outlets, the reports contain phrases most readers have subconsciously come to expect and accept:

> *Cause of death unknown, died of natural causes, no cause of death has been given, died after a brief illness, officials are still investigating cause of death*

No follow-up news stories, no questions from reporters, no curiosity.

But I am curious. I've made a career of being curious about trends, early indicators, statistics, demographics, metrics. That's what I do: draw useful patterns from as many sources of reliable information and data as I can get.

Essentially, my career on Wall Street was to make money for our clients by making the best predictions. Wall Street is mostly comprised of smart, hardworking men and women competing to create impressive returns for their clients. Like any of us, people on Wall Street want proof of concept before acting, and certainly before committing money to some idea. There is a Wall Street saying: "Be early, be right and be loud!" A common problem is that most people are hesitant to go against the grain of consensus; they prefer to ride trends everyone can see. But the big money in stock picking is made by identifying changes in trends before everyone sees them.

Investing huge amounts of money incentivizes a person to be correct, to discern the difference between mere claims and real possibilities. That required me to become a successful ambassador between two worlds: perception and reality.

On Wall Street, those who can figure out reality and see the near future before the herd's perception changes stand to make a lot of money. My brain is wired in this fashion, and I'm applying the same modality to the strange present moment in which public health officials are no longer curious about what causes unexpected death.

In addition to the young athletes I saw dying in 2021, I began to hear stories from friends whose loved ones had died suddenly. (Oddly, I hadn't heard such stories in 2020 when COVID was raging.) Before we look at the obvious change that occurred in 2021, I want to honor the healthy young people who died in their sleep, and whose deaths official agencies worked hard to ignore.

As you turn the following pages, ask yourself how often this happened while you were growing up. Reality is more persuasive than all the public health officials, Pharma executives and Pharma-sponsored talking heads on the cable news. We'll get into the statistical sciences in a moment, but likely by then you'll have already realized on your own that something unusual and terrible is happening.

Was a professional for nearly 20 years, played nearly 500 games, club and international level. Never ever was there one cardiac arrest—either in the crowd or a player. It's actually quite scary.

— Gary Dempsey, 2011 Footballer of the Year

DIED UNEXPECTEDLY IN
THEIR SLEEP...

9-year-old Utah boy dies unexpectedly in sleep, cause unknown

Posted at 8:56 PM, Mar 14, 2022 and last updated 2:57 PM, Mar 17, 2022

CONFIRM THIS STORY IS REAL

 NEW YORK POST LOG IN

UK mom dies in her sleep on flight with husband and two kids

August 8, 2022 | 8:58pm | Updated

CONFIRM THIS STORY IS REAL

UNION NEWS DAILY

Rahway varsity football player dies in his sleep

By JR Parachini on September 16, 2022

CONFIRM THIS STORY IS REAL

Daily Mail.com   News

'Beautiful and kind' mother-of-two, 23, dies suddenly in her sleep

PUBLISHED: 15:21 EDT, 19 November 2021 | UPDATED: 15:35 EDT, 19 November 2021

CONFIRM THIS STORY IS REAL

NEW YORK POST   LOG IN

Rep. Sean Casten says his 17-year-old daughter Gwen died peacefully in her sleep

June 15, 2022 | 8:03pm | Updated

CONFIRM THIS STORY IS REAL

EXPRESS

Golf starlet Elexis Brown, aged 13, dies in sleep

00:51, Sun, Aug 15, 2021 | UPDATED: 15:28, Sun, Aug 15, 2021

CONFIRM THIS STORY IS REAL

DIED UNEXPECTEDLY IN THEIR SLEEP...

NationalWorld

Will Jones: sudden death of dad, 26, found dead after 'falling asleep' leaves family 'devastated'

Monday, 20th June 2022, 3:25 pm

CONFIRM THIS STORY IS REAL

Mum, 26, 'fell asleep and didn't wake up' on car trip back from Wales

21:04, 31 May 2022 | UPDATED 07:19, 1 Jun 2022

Mirror

CONFIRM THIS STORY IS REAL

Daily Record

Tragic Scots student dies in his sleep on adventure holidays in French Alps

18:10, 10 JUN 2022 UPDATED 10:01, 12 JUN 2022

CONFIRM THIS STORY IS REAL

PADOVAOGGI

Found dead at a friend's house: in his twenties he falls asleep and never wakes up

August 17, 2022 01:02

CONFIRM THIS STORY IS REAL

ImolaOggi.it
Direttore Armando Manocchia

Sudden illness, 22-year-old girl dies in her sleep

August 10, 2022

CONFIRM THIS STORY IS REAL

Lucca*in*Diretta

The emerging painter Arianna Mora dies in her sleep at the age of 33

by Redazione - 03 August 2022 - 22:31

CONFIRM THIS STORY IS REAL

DAILY VOICE

Beloved Trenton Native Dies In Sleep, 28

08/19/2022 11:50 a.m.

CONFIRM THIS STORY IS REAL

NEW YORK POST

Michigan boy dies in his sleep three days after getting vaccine

By Joe Tacopino   July 5, 2021 | 2:45am | Updated

CONFIRM THIS STORY IS REAL

The Sun

MOTHER'S HEARTACHE My fit and healthy daughter, 31, died suddenly in her sleep – I never saw it coming

19:34, 30 May 2022 | Updated: 0:41, 31 May 2022

CONFIRM THIS STORY IS REAL

OBSERVATOR

Alessia Raiciu, a young basketball player from Bucharest, died in her sleep on the day she turned 18

By the Observator editorial team on 15.08.2022, 20:25

CONFIRM THIS STORY IS REAL

PADOVA OGGI

Fatal illness in sleep: Matteo dies at 27

April 29, 2022 9:05 am

CONFIRM THIS STORY IS REAL

PARALLELO

Drama in Rimini, Diego falls asleep on the sofa and never wakes up again: he died at just 19, promising football

 Published 6 months ago the March 3, 2022
From **Irene Viturri**

CONFIRM THIS STORY IS REAL

The first prospective core study with Pfizer vaccine ages 13 to 18 found that a large fraction, nearly half, were asymptomatic of myocarditis. That means a substantial number of young people in fact are sustaining heart damage and they don't know it, their parents don't know it, and the first manifestation of heart damage can be cardiac arrest. This can happen on the playing field with exertion—it also can happen during sleep—as well as the development of heart failure later on.

— Dr. Peter McCullough, Leading Cardiologist

AUSTRALIAN NATIONAL REVIEW

📅 July 17, 2021 👤 ANR News

Camilla Canepa, 18, London UK, Died After First Vaccine Dose

May 25 - Received Covid vaccine

June 3 - Checked into the emergency room suffering severe headache and light sensitivity

June 5 - returned to emergency room, paralyzed. Blood clot found between eye socket and brain, also bleeding in the brain, diagnosed with cavernous sinus thrombosis

June 10 - Died

CONFIRM THIS STORY IS REAL

prima VERONA

Died in his sleep at 27: it was his mother who found him lifeless in bed

CHRONICLE Legnago and lower , 12 July 2022 at 14:41

CONFIRM THIS STORY IS REAL

IL MATTINO.it

Ariano in shock for Michele who died in bed

Friday 31 December 2021 by *Monica De Benedetto*

CONFIRM THIS STORY IS REAL

KCCI 8 DES MOINES 💬 6

Drake honors former softball player's memory

Updated: 7:19 PM CDT Apr 23, 2022

CONFIRM THIS STORY IS REAL

EuroWeekly NEWS
NEWS AND VIEWS SINCE 1998

Heartbreak as 24-year-old Italian man dies suddenly and unexpectedly in Spain's Mallorca

By Matthew Roscoe · 02 September 2022 - 17:29

CONFIRM THIS STORY IS REAL

Former junior world champion John Paul dies aged 28

PUBLISHED MARCH 10, 2022

CONFIRM THIS STORY IS REAL

GoFundMe fundraiser launched for Ben Penrose

Updated August 21 2022 - 9:11pm, first published 9:09pm

CONFIRM THIS STORY IS REAL

Centennial High School football shocked by death of lineman Cesar Vazquez

Published 7:49 a.m. MT Aug. 4, 2022 | Updated 9:47 a.m. MT Aug. 5, 2022

CONFIRM THIS STORY IS REAL

San Lorenzo in mourning: Andrea Dorno, the soul of Atletico basketball and Communia, has died

March 24, 2022 4:21 pm

CONFIRM THIS STORY IS REAL

RUTHERFORD
CARLSTADT   EAST RUTHERFORD   + Follow

Beloved Sparta High School Grad, Johns Hopkins Research Technologist Dies Suddenly, 26

04/11/2022 12:05 p.m.

CONFIRM THIS STORY IS REAL

TCPalm.

Football player Nikolas Lawrynas, 17 and class of 2021 senior, dies

Published 11:12 a.m. ET May 3, 2021 | Updated 7:09 p.m. ET May 3, 2021

CONFIRM THIS STORY IS REAL

L'UNIONE SARDA *.it* World Italy Sardinia Sardinians in the World

THE TRAGEDY

Matteo Cannavera, manager and former tennis player from Cagliari, died in New York

★ **StarTribune**

Katie Novak, passionate fitness trainer, dies at 31

By Mara Klecker Star Tribune   FEBRUARY 3, 2022 — 3:21PM

⑤ WDTV

Rising BU junior, student athlete dies

Published: Aug. 16, 2022 at 7:40 PM PDT

Caitlyn Gable

N NEWS SPORT REGION SMARTER LIVING BILLIE PODC

Youth trainer and father of four Jeffrey (32) unexpectedly died

Olivier Simons
Tuesday, February 22, 2022 at 10:55 PM

L'UNIONE SARDA *.it*

Alessandro Tedde, struck down by an illness at the age of 26

April 14, 2022 at 4:37 pm

≡   **LaVoz**

The young Alejandro Candela dies suddenly, one of the promises of swimming in Lanzarote

LAVOZDELANZAROTE.COM   JUNE 19, 2018 (21:31 CET)

itv   Sign In ⊖ ≡

Hundreds of tributes pour in for 18-year-old Bolton rugby player Cameron Milton

GRANADA | SPORT | BOLTON | ⏱ Thursday 31 March 2022 at 9:43am

≡
MENU   ouest france ⊙

Handball. Former captain of Angers Sco, Gillen Lusson has died

Published on27/12/2021 at 07:02

≡ MENU   **THE IRISH NEWS**   Q

Family left 'completely crushed' following sudden death of father-of-two

07 September, 2022 03:00

interia SPORT   ✉ ≡

Hockey. Former MMKS Podhale Nowy Trag player Krzysztof Steliga died

Tuesday, September 21, 2021 (11:08 AM)

⑤ **SPORTBUZZER**

July 19, 2021 / 5:34 p.m

27-year-old footballer dies after football tournament

LE PROGRÈS

Former Choralien Arthur Zuccolini dies at 29

By Le Progrès - Jul 15, 2021, 2:29 | updated on Jul 15, 2021 at 15:46 - Reading time: 1 min

Daily **Mail**
.com

Cryptocurrency co-founder Tiantian Kullander dies 'unexpectedly' aged 30

BY RACHAEL BUNYAN FOR MAILONLINE
PUBLISHED: 05:57 EST, 28 November 2022 | UPDATED: 10:06 EST, 28 November 2022

Brad William Henke, 'Orange Is the New Black' Actor and Former NFL Star, Dead

By Glenn Garner   Published on December 1, 2022 08:53 PM

The Philadelphia Inquirer   SIGN IN /

PHILADELPHIA

Frank Tartaglia, South Philly filmmaker and co-owner of Connie's Ric Rac club, dies at 45

by Mike Newall
Published Nov 28, 2022

la Repubblica

A student dies at the age of 17 at home from a sudden illness

NOVEMBER 14, 2022   UPDATED AT3:59PM

Tayler Marie Woolston

You've just seen a very small sampling of sudden deaths during 2021 / 2022

A sampling of many hundreds more appears in the Compendium on Page 145

THE SAD NEW NORMAL

Before we move to the statistical science, I want to extend my condolences to the many people affected by these terrible losses. Sudden unexplainable deaths profoundly hurt families, extended family, friends and community. We can all accept accidents —the leading cause of death in the young— but healthy teens dying in their sleep, or collapsing and dying on the spot during sports? Watching my son's soccer game, I can't even imagine it. And rather than sincere investigation, there has been a concerted effort to normalize these horrible events.

Among the examples of normalization is the promotion of Sudden Adult Death Syndrome, or SADS. That's "when a person under the age of 40 years old suffers a sudden death without a known cause." In other words, when circumstances and even autopsy cannot identify a cause of death, let's label that mystery SADS. Let's pretend SADS is a syndrome or a health condition when it's actually "an umbrella term to describe unexpected deaths in young people." It's a term, not an explanation – used when there is no explanation. It's a category of death, not a cause of death. A person can't be diagnosed with SADS, and SADS cannot kill people; it can only distract people.

Daily Mail

Healthy young people are dying suddenly and unexpectedly from a mysterious syndrome - as doctors seek answers through a new national register

- People aged under the age of 40 being urged to go and get their hearts checked
- May potentially be at risk of having Sudden Adult Death Syndrome (SADS)
- SADS is an 'umbrella term to describe unexpected deaths in young people'

By TOM HEATON FOR DAILY MAIL AUSTRALIA
PUBLISHED: 02:05 EST, 8 June 2022 | UPDATED: 02:32 EST, 8 June 2022

EuroWeekly NEWS
NEWS AND VIEWS SINCE 1998

World News

Doctors baffled by Sudden Adult Death Syndrome (SADS) in healthy young people

By Matthew Roscoe · 08 June 2022 · 11:18

Health Desk

BACK

What is Sudden Adult Death Syndrome?

by **Health Desk** | Published on June 7, 2022 –
Updated on June 7, 2022 | Explainer

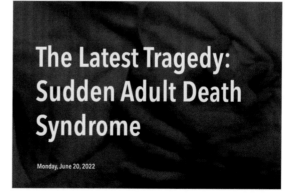

The Latest Tragedy:
Sudden Adult Death Syndrome

Monday, June 20, 2022

By April of 2021 I could see and begin to test the most compelling thesis about what was causing the shocking spike in sudden deaths of healthy young people. Trained to observe shifts in trends, it was clear to me that one big change could not be ignored: mass administration of new pharmaceutical products, experimental products using novel technology with only a 28-day clinical trial testing period. Under Operation Warp Speed, these products were quickly deemed "safe and effective" by the FDA. By April 2021, about 50% of eligible Americans had been injected with the new products.

There are simple facts that can be assessed and analyzed. For example, the number of Americans who took the vaccine products in 2021 can be laid alongside an increase in sudden deaths.

In public health, if evidence indicates a connection between a new Pharma product and increased death, our government historically stopped promoting and administering the product.

Instead, we saw history's biggest advertising and propaganda campaign, including the use of mandate and intimidation to force people to take the new products.

I had been here many times before in my stock picking career. The Dotcom bubble had fanatics who proclaimed it a "new paradigm." I thought they were crazy and parlayed my contrary bet into substantial success. Five years later, the battle cry was "home prices never go down!" and again my teammates and I managed to steer our fund through that fraud, gaining tremendous market share as a result.

In April of 2021, we all met a new battle cry: "Safe & Effective!"

But I saw the trend of sudden death where none had been before.

VACCINE MANDATES & ALL-CAUSE MORTALITY

In Spring and Summer of 2021, we all watched President Biden famously declare that we were experiencing "a pandemic of the unvaccinated." The CDC, Anthony Fauci's NIH, and the medical establishment were touting the COVID vaccines as 100% effective against getting or spreading the virus. Though we all hoped for a successful vaccine, these claims were proven false. We all now know that COVID vaccine products do not stop infection, do not stop disease, and most important from a public policy point of view, do not stop transmission.

Simple logic dictates that if you are vaccinated you shouldn't fear an unvaccinated person. But there was a hysteria pushed by our government that led to mandates for the general public, following the already draconian mandates forced upon hospital workers and first responders in early summer 2021.

Soon, people would need proof of vaccination to use restaurants, gyms, concerts, sporting events, attend college, keep a job, etc. Remember, this was for a virus that had a survival rate of 99.8% and mostly affected old people who were already sick with one or more fatal comorbidities. Though the survival rate isn't widely reported even today, it was already known right near the start of the pandemic: Most people had a very low risk of death from COVID. Way back in March of 2020, the facts from Italy showed that nearly 100% of deaths attributed to COVID were in people already ill with more than one fatal comorbidity.

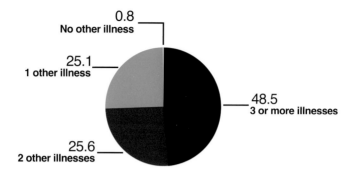

Italy Coronavirus Deaths
By prior illness (%)

0.8
No other illness

25.1
1 other illness

48.5
3 or more illnesses

25.6
2 other illnesses

Source: ISS Italy National Health Institute, March 17 sample

If my thesis about the risks of the new vaccine products was right, then All-Cause Mortality would explode when mandates were implemented.

I also formed the professional opinion that these excess deaths would soon show up in insurance company and funeral home results. Insurance companies would incur abnormal losses and funeral home companies would experience growth beyond their expectations.

Through Summer 2021, corporations began their vaccine mandates; Goldman Sachs and Morgan Stanley, two of the most revered investment banks on Wall Street, led the way, announcing vaccine mandates in August 2021. Most of corporate America followed this lead. Then in September, President Biden signed an executive order mandating COVID vaccination for all companies with over 100 employees. The fear campaign by the CDC, NIH and the Biden administration was aided by a complicit news media that's long been captured by Big Pharma.

The vaccination rate, particularly among working people, rose extremely fast in a short amount of time. The data would soon confirm that **being employed in 2021 was actually detrimental to your health**.

By the fall-winter of 2021 the threatened mandates were implemented by businesses all over America. We all saw the mainstream media demonize the unvaccinated, the effect of which was to tear families and friends apart as many who took the vaccine refused to mingle with the "infectious" unvaccinated. The country was encouraged to forget the traditional understanding of the word vaccine: an intervention that would prevent nearly everyone from getting the infection from an unvaccinated person. Since this vaccine didn't stop infection or transmission, the old definition of vaccine was becoming less applicable by the day.

As the news reported an alarming number of what they euphemistically called "breakthrough cases," it soon became so obvious that the CDC decided to quietly change the definition of vaccine. On September 1, 2021, the definition was changed from inoculants that "produce immunity to a specific disease" to "a preparation that is used to stimulate the body's immune response against diseases."

The definition change turned these new COVID products into therapeutics, not vaccines.

<u>Note</u>: Throughout this book, the word vaccine is used for ease of reading – not because these COVID products are vaccines according to any traditional definition.

The CDC, NIH and even President Biden himself told everyone that if you take the vaccine you are protected from COVID and cannot spread COVID. This is not a little lie; it is one to keep in mind as you view the All-Cause Mortality data in this book.

Unfortunately, perception was woefully far from reality. The mainstream media didn't bother to mention the definition change, continued to spout "safe and effective," and continued to demonize the unvaccinated.

As everyone saw that Omicron spread more easily, the perception that the vaccine prevented people from getting COVID was shattered. Literally months after it was obvious to everyone, the corporate mainstream media began to quietly report that the COVID vaccines don't stop transmission. Still today, many people haven't absorbed this fact, a fact known to Federal public health officials before mass vaccination even began.

The narrative then shifted from effective to "prevents serious hospitalization and death," a claim that to this day has never been proven, a claim that is more marketing than science.

With Omicron, nature had accomplished what no amount of logic or reasoning or rational debate could do: It changed perception to match the reality that these vaccines were close to useless. CDC Director Walensky: 4 shots plus the new bivalent booster, and gets COVID. Dr Fauci: 4 shots and gets COVID. Twice. President Biden: 4 shots and gets COVID. Twice. Jill Biden: 4 shots and gets COVID. Twice. Canada's Prime Minster: Fully boosted and gets COVID. Twice. Even the CEO of Pfizer: Fully vaccinated and gets COVID. Twice.

When President Biden assured Americans "You're not going to get COVID if you have these vaccinations," people quickly saw that wasn't accurate. And when President Biden explained his mandate for healthcare workers, he told Americans they could have "certainty" that hospital staffs are "protected from COVID and cannot spread it to you." But Americans saw first-hand that was false.

Naturally, booster uptake lessened, and unnaturally, the government and their media allies continued to push this vaccine product, as they still do today. Unfortunately, the damage done by this public health disaster was about to show up in the meta data of the CDC and similar agencies worldwide, as well as in insurance company and funeral home earnings reports.

It should be noted that the Government's Vaccine Adverse Event Reporting System (VAERS), established decades ago to detect safety signals from vaccine products, was dramatically revealing problems with the vaccines.

Media and government leaders tried to dismiss VAERS out of hand, implying that anti-vaxxers were using it to submit false claims. But VAERS has nearly 1.5 million reports of adverse reactions and deaths associated with the COVID vaccines, most entered by medical professionals, emergency room staff, etc. The idea that significant numbers of these reports are fake is ludicrous. Aside from it being a Federal offense to file a false VAERS report, it's impenetrably difficult to enter a VAERS report. Doubting readers can try it right now.

Those rare medical professionals who pointed out the reality of vaccine injuries and deaths were demonized as conspiracy theorists and threatened with loss of their jobs and licenses. And though the VAERS data was ignored, the bodies were starting to pile up, and evidence within other databases would soon be impossible to refute.

People with myocarditis are at lifetime increased risk of cardiac complications. This can have profound consequences... typically told to limit activity for several months, sometimes longer. This means no sports. Some kids are told not to carry books to school.

— Dr. Venk Murthy
Johns Hopkins-trained Cardiologist

VAERS
THE U.S. GOVERNMENT'S VACCINE ADVERSE EVENT REPORTING SYSTEM

From a study of VAERS commissioned by the U.S. Government:

Low reporting rates preclude or slow the identification of problem drugs and vaccines that endanger public health. New surveillance methods for drug and vaccine adverse effects are needed...fewer than 1% of vaccine adverse events are reported.

Even with these reporting deficits, there have been more adverse reactions and deaths reported to VAERS for the COVID vaccines than all other vaccines combined, over 32 years.

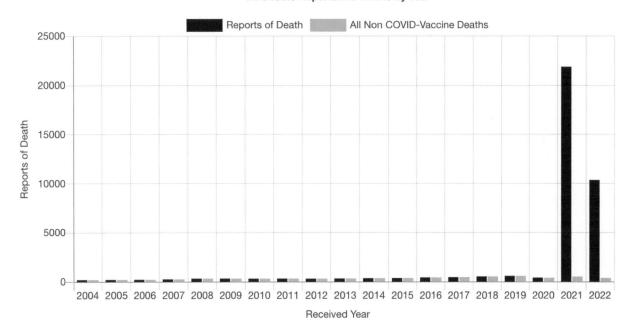

Thousands of deaths have occurred the day of Covid vaccination, or within one or two days.

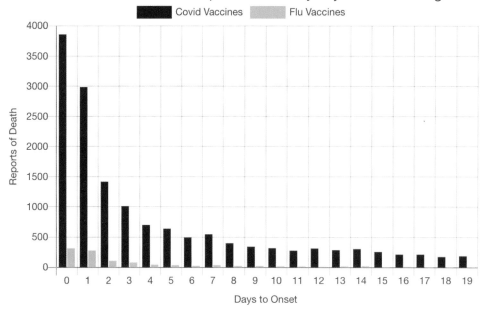

VAERS COVID/FLU Vaccine Reported Deaths by Days to Onset All Ages

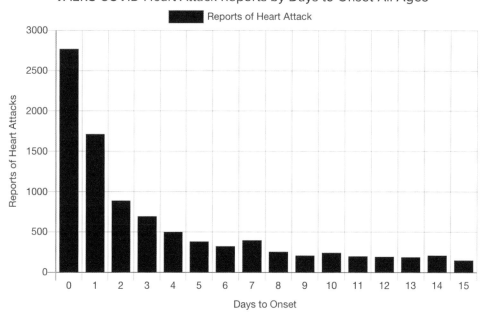

VAERS COVID Heart Attack Reports by Days to Onset-All Ages

All Myo/Pericarditis Reported to VAERS by Year (all vaccines)

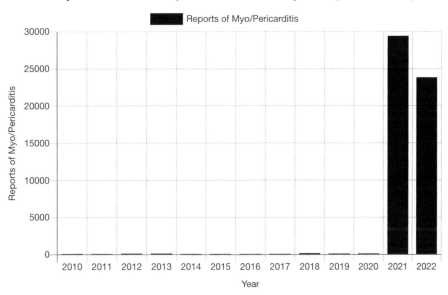

All Myo/Pericarditis Reported to VAERS Post COVID Vaccine by Dose

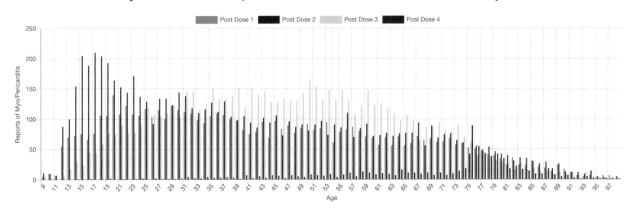

Reproductive Adverse Events

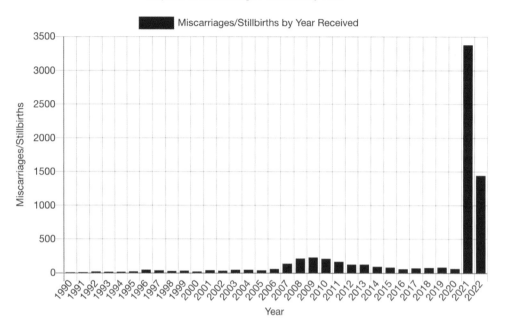

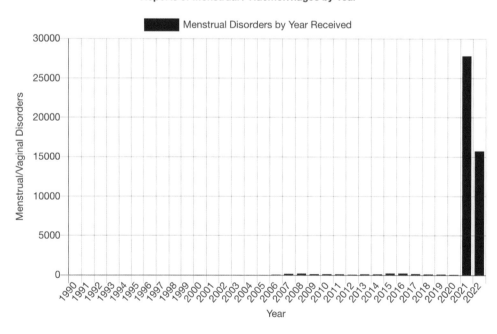

1,481,226 COVID Reports Through December 09, 2022

32,828 DEATHS	**186,098** HOSPITALIZATIONS	**141,417** URGENT CARE
216,621 DOCTOR VISITS	**10,240** ANAPHYLAXIS	**16,481** BELL'S PALSY

4,623 Miscarriages	15,801 Heart Attacks	35,820 Myocarditis/ Pericarditis	61,065 Permanently Disabled
8,301 Thrombocytopenia/ Low Platelet	35,425 Life Threatening	41,785 Severe Allergic Reaction	15,200 Shingles

A Sample from Among Tens of Thousands of VAERS Reports

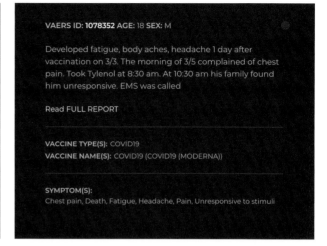

VAERS ID: 1054160 AGE: 36 **SEX:** M

2/12/2021 woke up with sore arm and back. 2/13/2021 woke up with headache around 1am. Headache and nausea all morning. Mid-late afternoon started having seizures. Admitted to Hospital 2/15/2021 expired.

Read FULL REPORT

VACCINE TYPE(S): COVID19
VACCINE NAME(S): COVID19 (COVID19 (MODERNA))

SYMPTOM(S):
Back pain, Death, Headache, Nausea, Pain in extremity, Seizure

VAERS ID: 1078352 AGE: 18 **SEX:** M

Developed fatigue, body aches, headache 1 day after vaccination on 3/3. The morning of 3/5 complained of chest pain. Took Tylenol at 8:30 am. At 10:30 am his family found him unresponsive. EMS was called

Read FULL REPORT

VACCINE TYPE(S): COVID19
VACCINE NAME(S): COVID19 (COVID19 (MODERNA))

SYMPTOM(S):
Chest pain, Death, Fatigue, Headache, Pain, Unresponsive to stimuli

VAERS ID: 1243791 AGE: 21 **SEX:** M

Per the father, the deceased received his first shot of Moderna vaccine on Saturday, 4/10/2021 at a local church. He did not work on 4/11/2021. Worked on 4/12/2021. The deceased was found dead at 6:43

Read FULL REPORT

VACCINE TYPE(S): COVID19
VACCINE NAME(S): COVID19 (COVID19 (MODERNA))

SYMPTOM(S): Death

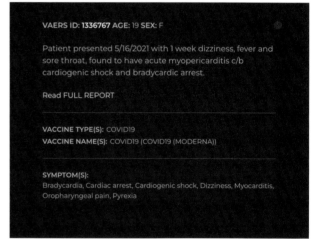

VAERS ID: 1336767 AGE: 19 **SEX:** F

Patient presented 5/16/2021 with 1 week dizziness, fever and sore throat, found to have acute myopericarditis c/b cardiogenic shock and bradycardic arrest.

Read FULL REPORT

VACCINE TYPE(S): COVID19
VACCINE NAME(S): COVID19 (COVID19 (MODERNA))

SYMPTOM(S):
Bradycardia, Cardiac arrest, Cardiogenic shock, Dizziness, Myocarditis, Oropharyngeal pain, Pyrexia

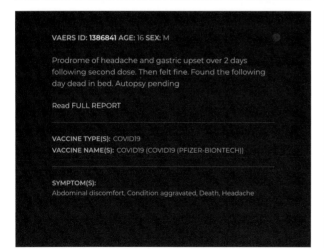

VAERS ID: 1386841 AGE: 16 **SEX:** M

Prodrome of headache and gastric upset over 2 days following second dose. Then felt fine. Found the following day dead in bed. Autopsy pending

Read FULL REPORT

VACCINE TYPE(S): COVID19
VACCINE NAME(S): COVID19 (COVID19 (PFIZER-BIONTECH))

SYMPTOM(S):
Abdominal discomfort, Condition aggravated, Death, Headache

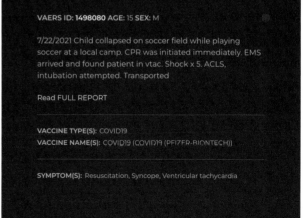

VAERS ID: 1498080 AGE: 15 **SEX:** M

7/22/2021 Child collapsed on soccer field while playing soccer at a local camp. CPR was initiated immediately. EMS arrived and found patient in vtac. Shock x 5. ACLS, intubation attempted. Transported

Read FULL REPORT

VACCINE TYPE(S): COVID19
VACCINE NAME(S): COVID19 (COVID19 (PFIZER-BIONTECH))

SYMPTOM(S): Resuscitation, Syncope, Ventricular tachycardia

Bill Gates 2020:

The vaccine key goal is to stop the transmission, to get immunity levels up so that you get almost no infection going on whatsoever.

Bill Gates 2021:

We didn't get vaccines that stop transmission. We need a new way of doing the vaccines.

YOUNG MEDICAL PRACTITIONERS...

FOX3now

20-Year-Old Kansas Girl Regan Laine Lewis Dies of Cardiac Arrest Within 24 Hours of Her COVID Vaccine

October 2, 2022   FOX3 Now Staff   COVID Vaccine, Kansas, Regan Laine Lewi

CONFIRM THIS STORY IS REAL

28-Year-Old Mother Has Fatal Stroke 5 Days After COVID Vaccine

February 12, 2021

CONFIRM THIS STORY IS REAL

Daily Mail.com

Physical therapist, 28, working at a senior living facility dies two days after getting the COVID-19 vaccine

PUBLISHED: 00:42 EDT, 13 March 2021 | UPDATED: 03:39 EDT, 13 March 2021

CONFIRM THIS STORY IS REAL

CTV NEWS
TORONTO   ● LIVE

TORONTO | News

Ontario paramedic dies suddenly while on jet ski at U.S. cottage, family says

Ria Vannoort, left, is seen with her six-year-old daughter. (Supplied)

Sean Davidson
CTVNewsToronto.ca Managing Digital Producer
 Follow | Contact

Published Aug. 23, 2022 2:03 p.m. PDT

Police said when they arrived they tried to resuscitate a 32-year-old woman, who was later identified as Ria Vannoort.

"Initial investigation by deputies at the scene revealed that Vannoort, who was attending a party at the residence, was operating a personal watercraft just offshore," police said. "Attendees of the party noticed that Vannoort had become dismounted from the watercraft and was floating in the lake. Vannoort was then loaded onto a second vessel and brought back to shore by the other attendees."

░CBC | MENU ˅   Q Search   ⎙ Sign In

Doctor at McMaster Children's Hospital dies after competing in Toronto Triathlon Festival

Posted: Jul 30, 2022 6:00 AM ET | Last Updated: July 30

CONFIRM THIS STORY IS REAL

CONFIRM THIS STORY IS REAL

Waldport Clinic doctor Stephanie Bosch dies "suddenly and unexpectedly"

OCTOBER 14, 2021

CONFIRM THIS STORY IS REAL

IL GAZZETTINO.it

36-year-old doctor struck down by heart attack while jogging

Sunday 17 October 2021 by **Cesare Arcolini**

CONFIRM THIS STORY IS REAL

Canadian Chief of Emergency Medicine, Olympic Sailor, and Marathon Runner Dies Suddenly While on a Run

By Jim Hoft
Published July 17, 2022 at 3:11pm

CONFIRM THIS STORY IS REAL

We report the truth— and leave the Russia-Collusion fairy tale to the Conspiracy media

26-Year-Old Neurosurgery Resident Dies in July, Making Seven Canadian Doctors to Die in Two Weeks

Published August 15, 2022 at 12:45pm

CONFIRM THIS STORY IS REAL

INDEPENDENT   Subscribe

NEWS   SPORTS   VOICES   CULTURE   LIFESTYLE   TRAV

Lifestyle > Health & Families

Family of nurse who died weeks after giving birth share blood clot symptoms to warn others

Samantha Crosbie of Surrey tragically died aged just 32, leaving he
three children with one picture of them all altogether

Louise Lazell • 5 days ago

CONFIRM THIS STORY IS REAL

WalesOnline

Dad, husband and child psychologist died suddenly - aged 32

07:43, 5 FEB 2021

CONFIRM THIS STORY IS REAL

This would not be the first time if it happened that a vaccine that looked good in initial safety actually made people worse... the respiratory syncytial virus vaccine in children which paradoxically made the children worse.* One of the HIV vaccines that we tested several years ago actually made individuals more likely to get infected.

So you can't just go out there and give it.

— Anthony Fauci on Vaccines March 2020

* 80% of the children given the shot were hospitalized with severe respiratory disease

NAPOLITODAY

Remigio Bova died, the nurse and referee was 30 years old

October 23, 2021 6:31 pm

CONFIRM THIS STORY IS REAL

Daily Mail.com

Portuguese health worker, 41, dies two days after getting the Pfizer covid vaccine as her father says he 'wants answers'

PUBLISHED: 10:20 EDT, 4 January 2021 | UPDATED: 05:34 EDT, 11 January 2021

CONFIRM THIS STORY IS REAL

Daily Record

Superfit Scots doctor dies after having heart attack while swimming in loch

14:35, 2 SEP 2022   UPDATED 20:45, 2 SEP 2022

CONFIRM THIS STORY IS REAL

People   SUBSCRIBE

Olivia Newton-John Mourns the 'Sudden' Death of Her Cancer Nurse: 'My Heart Is Still in Shock'

By Katie Campione   Published on April 28, 2021 11:53 PM

CONFIRM THIS STORY IS REAL

GATEWAY PUNDIT
We report the truth — and leave the Russia-Collusion fairy tale to the Conspiracy media

43-Year-Old Medical Doctor, Author and Editor Dies Suddenly After Seizure

By Kristinn Taylor
Published August 12, 2022 at 1:35pm
778 Comments

CONFIRM THIS STORY IS REAL

NEW YORK POST

David Reichman, runner who died in Brooklyn Half Marathon, was NYC therapist

By Tina Moore, Kevin Sheehan and Jorge Fitz-Gibbon   May 22, 2022 | 2:27pm | Updated

CONFIRM THIS STORY IS REAL

More Young Canadian Doctors Who Died Suddenly Since March 2021

Died: July 23, 2022
Dr. Shahriar Jalali Mazlouman
Age: 44
Melville, SK
Family physician
Died swimming in a public pool

Died: July 30, 2022
Dr. Michael Mthandazo
Age: 40s
Vernon, BC
Family physician
Died swimming in a river

Died: July 19, 2022
Dr. Jakub T. Sawicki
Age: 40s
Mississauga, ON
Family physician
Gastric cancer, Stage 4, < 1 year

Died: Dec 23, 2021
Dr. Neil Singh Dhalla
Age: 48
Toronto, ON
Family physician, Activa Clinics
Died in sleep 4 days post 3rd jab

Died: May 16, 2022
Dr. Joshua Raj Kotaro Yoneda
Age: 27
Kamloops, BC
Medical student, UBC
Rare spinal cord glioma, < 1 year

Died: May 14, 2022
Dr. David Laverdiere
Age: 48
Chicoutimi, QC
Internal medicine, Respirology
Died unexpectedly

Died: July 17, 2022
Dr. Lorne E. Segall
Age: 49
Mississauga, ON
ENT specialist
Lung ca, Stage 4, < 1 year

Died: July 13, 2022
Dr. Baharan Behzadizad
Age: 40s
Newfoundland
Family physician
Died in her sleep

Died: Dec 21, 2021
Dr. Cintia Vontobel Padoin
Age: 44
North Bay, ON
Psychiatrist
Malignant melanoma, < 1 year

Died: May 19, 2022
Dr. Wilson Idami
Age: 54
Aurora, ON
Family physician
Died unexpectedly

Died: March 20, 2021
Dr. Inderjit Andy Jassal
Age: 42
Surrey, BC
Family physician
Died unexpectedly, heart attack

Died: June 6, 2022
Dr. Arran Lamont
Age: 38
Stony Plain, AB
Veterinary medicine, surgery
Died unexpectedly

Died: March 14, 2022
Dr. Bradley James Harris
Age: 49
mox, BC
amily physician
ied while running

Died: Feb 13, 2022
Dr. Michael Stefanos
Age: 50
Mississauga, ON
Radiologist
Died in his sleep, heart attack

Died: Nov 08, 2021
Dr. Sohrab Lutchmedial
Age: 52
Saint John, NB
Cardiologist
Died in sleep 2 wk post 3rd jab

Died: June 25, 2021
Dr. Ainsley Moore
Age: 57
Hamilton, ON
Family physician
Died unexpectedly, heart attack

Died: June 20, 2021
Dr. Catherine Yanchula
Age: 56
Windsor, ON
Family physician
Died unexpectedly

Died: June 19, 2021
Dr. Dick Au
Age: 53
Edmonton, AB
Internist, Geriatrician
Died of "sudden vascular event"

In the first week of January 2022 OneAmerica CEO Scott Davison made comments to a Commerce meeting that were soon picked up by some media outlets:

> *"We are seeing, right now, the highest death rates we have seen in the history of this business – not just at OneAmerica. The data is consistent across every player in that business."*

Davison said the increase in deaths represented "huge, huge numbers," and that it wasn't elderly people who were dying, but "primarily working-age people 18 to 64" who were employees of companies with group life insurance plans through OneAmerica.

"And what we saw just in third quarter, we're seeing it continue into fourth quarter, is that death rates are up **40%** over what they were pre-pandemic," he said.

"Just to give you an idea of how bad that is, a three-sigma or a one-in-200-year catastrophe would be 10% increase over pre-pandemic," he said. "So 40% is just unheard of."

He added that most of the claims for deaths being filed were not classified as COVID-19 deaths.

At the same time, the company was seeing an "uptick" in disability claims, at first short-term disability, and then a new increase in long-term disability claims.

MARKETS

Rise in Non-Covid-19 Deaths Hits Life Insurers

By *Leslie Scism* [Follow]

Updated Feb. 23, 2022 5:36 am ET

THE HILL

Well-Being > Longevity

'Huge, huge numbers:' insurance group sees death rates up 40 percent over pre-pandemic levels

By Shirin Ali | Jan. 7, 2022 | Jan. 07, 2022

'Just Unheard Of': Another Insurance CEO Admits Unexplained Deaths Are Up 40% Among Working People

 Fact checked

◷ July 4, 2022   ⚇ Baxter Dmitry   ☒ News, US   💬 11 Comments

Why Are All-Cause Excess Deaths in the Under-45s So Much Higher This Year Than at the Height of the Pandemic?

BY **NICK RENDELL**   15 AUGUST 2022 2:00 PM

↱ SHARE

Excess deaths in 2022 'incredibly high' in Australia

The Australian government should be urgently investigating the "incredibly high" 13 per cent excess death rate in 2022, the country's peak actuarial body says.

 Frank Chung

 @franks_chung ◷ **6 min read** December 8, 2022 - 9:42AM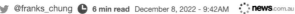

Seeing this strong corroboration for my concerns, I started to build a team that could carefully study each element of the alarming situation, starting with Josh Stirling, a former #1 ranked Institutional Investor and Wall Street Insurance Analyst who worked for Sanford C. Bernstein Research. Josh Stirling turned our focus on what's called the loss ratio of the group life and disability divisions of insurance companies. The change was significant because group life is a very stable business that is highly profitable for insurance companies. It's basically a death benefit and disability offered to mid-level employees when they join a corporation. The death benefit is not a big dollar amount and isn't expected to be paid out since statistically speaking, working age people don't die in large numbers; they are healthy and employed with good jobs.

The stunning data we found confirmed what the One America CEO saw in January 2022: The Fourth Quarter results from some of the major insurers saw a range of increase in their loss ratio of between 25% and 45% from the 2019 baseline levels. And the losses continued to rise. Many of the CEOs blamed this huge increase on COVID and a strange new term that someone coined: "indirect COVID."

Disability also saw a marked increase and continues to climb today. Remember, these are working age people who were not as a group much affected by COVID in 2020. After mass vaccination, however, they suddenly experienced a huge uptick in excess mortality. Simple deductive reasoning was clear: The only things that changed from 2020 to 2021 were the mandates and increasing mass vaccination. By all accounts, if the vaccines worked well, deaths and disabilities should have declined – but that's not what we saw happening.

Josh Stirling next studied the CDC excess mortality data. Though CDC data is poorly managed and expressed (not segmented by age, for example), the chart below is damning.

US data shows excess mortality has not fallen through the pandemic, not even after lockdowns, viral mutation to less severe strains, mass vaccination, herd immunity from prior infection, and advances in COVID treatment.

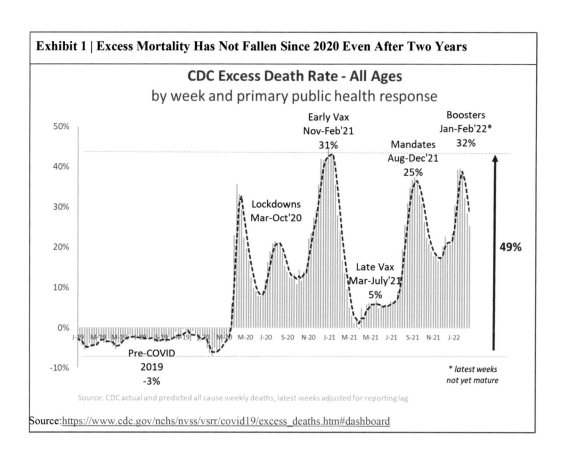

Exhibit 1 | Excess Mortality Has Not Fallen Since 2020 Even After Two Years

CDC Excess Death Rate - All Ages
by week and primary public health response

Source: CDC actual and predicted all cause weekly deaths, latest weeks adjusted for reporting lag

Source: https://www.cdc.gov/nchs/nvss/vsrr/covid19/excess_deaths.htm#dashboard

Drawing upon his own actuarial training and data he was able to collect from CDC's website, Josh developed pre-COVID excess mortality charts for each age group. What he found was simply stunning:

Millennials (ages 25-44) saw an acceleration of excess mortality into the second half of 2021 to new all-time highs, a stunning 84% above baseline.

Federal public health officials and media had developed ways to ignore and explain away the increase in All-Cause Mortality: It must be more suicides, or overdoses or missed cancer screening during lockdowns.

But the rate of change in Fall 2021 was particularly striking as it coincided with the corporate mandates – and it simply wasn't statistically possible that suicides, overdoses and deaths from delayed treatment of rapid-onset fatal cancers all spiked in that very same 3-month period. The only thing that had changed was mass vaccination forced upon the millennial generation via government and corporate mandates.

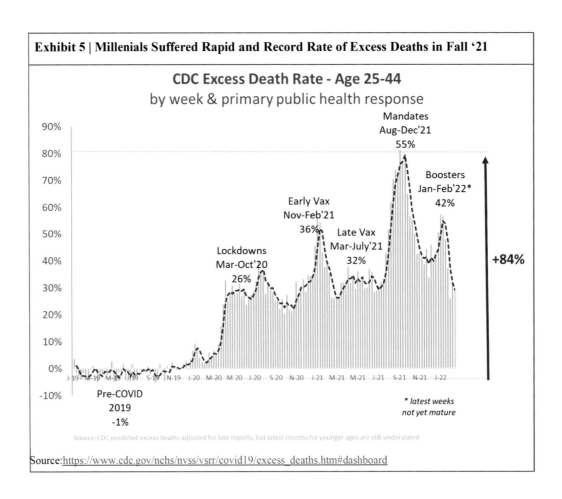

Exhibit 5 | Millenials Suffered Rapid and Record Rate of Excess Deaths in Fall '21

CDC Excess Death Rate - Age 25-44
by week & primary public health response

Source:https://www.cdc.gov/nchs/nvss/vsrr/covid19/excess_deaths.htm#dashboard

Josh Stirling's other critical discovery was the mix shift from old to young that occurred from 2020 to 2021 (next chart). Further analysis resulted in another series of revelations: In 2020 there were 126,000 excess deaths under the age of 65, or approximately 21%. In Year 2 of the Pandemic, there were 181,000 excess deaths of people under the age of 65, or approximately 35%. But the millennials saw the most enormous increase: 45%, from 42,000 excess deaths to 61,000 excess deaths.

It would be hard to explain the mix shift in Year 2 of the pandemic as being due to COVID because the mutating strains were already becoming less virulent, and we knew that the virus affected mostly older people with pre-existing fatal comorbidities.

In finance and stock picking, there are two kinds of mix shifts: favorable to earnings and adverse to earnings. A mix shift of deaths from old to young is beyond just adverse; it is tragic. Younger working age folks at peak earnings and societal contribution are dying at a faster rate compared to the older generation. Obviously, this is a terribly adverse mix shift, a group of deaths ignored by the media and Government – in stark contrast to their enthusiastic attention to death tolls in 2020.

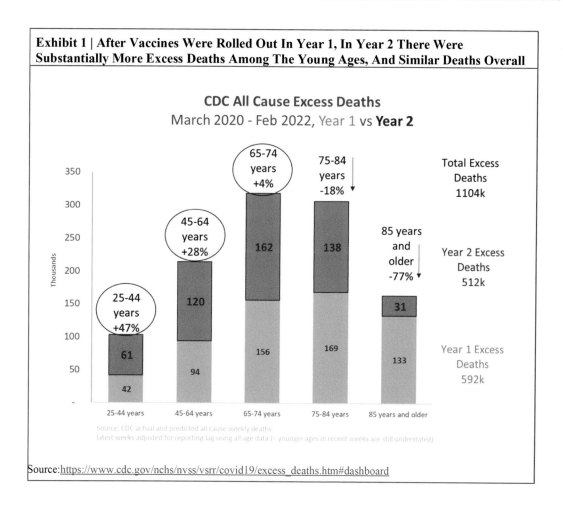

Exhibit 1 | After Vaccines Were Rolled Out In Year 1, In Year 2 There Were Substantially More Excess Deaths Among The Young Ages, And Similar Deaths Overall

Source:https://www.cdc.gov/nchs/nvss/vsrr/covid19/excess_deaths.htm#dashboard

How can public health officials explain these absolute increases in deaths among working age Americans? Did the virus suddenly decide to only target younger folks disproportionately? The only thing, and the most dramatic thing that changed in Year 2 was mass vaccination.

FAMOUS YOUNG PEOPLE...

Joyce Culla Death Cause Was Ruptured Brain Aneurysm Age 24, TikTok Star Passed Away

April 2, 2022 03:39 PM

CONFIRM THIS STORY IS REAL

DEADLINE

Charlbi Dean Dies Of Sudden Illness: Star Of Palme d'Or Winner 'Triangle Of Sadness' Was 32

By Greg Evans, Erik Pedersen
August 30, 2022 9:33am

CONFIRM THIS STORY IS REAL

METRO

Coronation Street actor died of 'sudden illness' while abroad

Thursday 24 Feb 2022 3:47 pm

CONFIRM THIS STORY IS REAL

Daily Mail.com

Neighbours star Miranda Fryer dies suddenly aged 34: Former actress who played Sky Mangel as a child 'went to sleep and never woke up', her devastated family reveal

PUBLISHED: 22:49 EDT, 10 January 2022 | UPDATED: 01:36 EDT, 11 January 2022

CONFIRM THIS STORY IS REAL

The Guardian
News website of the year

BBC presenter Lisa Shaw died of Covid vaccine complications, coroner finds

Shaw died of a brain haemorrhage caused by a blood clot three weeks after her first AstraZeneca dose

Thu 26 Aug 2021 08.05 EDT

CONFIRM THIS STORY IS REAL

The Indian EXPRESS

TV actor Sidharth Shukla passes away at 40

New Delhi | Updated: September 4, 2021 10:39:09 am

CONFIRM THIS STORY IS REAL

I will not get the third shot. I will not. Pfizer me once, no shame. Pfizer me twice, shame on Covid. Pfizer me three times, shame on you. You want me to get a third shot? What's next? A fifth shot? No, thank you.

— Comedian Nick Nemeroff in video he posted

Teen equestrian star Cienna Knowles hospitalised with blood clots after Pfizer vaccine

October 26, 2021 - 4:37PM

A "super healthy" teen equestrian competitor who was hospitalised last week with blood clots in her chest has blamed the potentially "life-changing" injuries on the Pfizer vaccine.

She was taken to Gosford Hospital where scans discovered blood clots in her legs, stomach and lungs.

"Crazy how quickly I went from a super healthy 19-year-old kid who's never had any form of health issues ever – working a full-time job, training and riding horses every day – to having it all taken away from me after my second Pfizer vaccination," Ms Knowles wrote on Facebook over the weekend.

"I wish I had never gotten it and I could have my healthy body back."

TMZ   NEWS   SPORTS   HIP HOP   WATCH   PHOTOS

'RHOA' Star NeNe Leakes 23-Year-Old Son Brentt Suffers Heart Attack and Stroke

NENE LEAKES
23-YEAR-OLD SON SUFFERS STROKE & HEART FAILURE

EXCLUSIVE   f 25.8K   🐦   10/10/2022 8:53 AM PT

THE U.S. Sun   HOME   NEWS   ENTERTAINMENT   LIFESTYLE   MONEY   HEALTH

BREAKING

Horror as radio DJ dies while on-air

Oct 24 2022

A RADIO host has tragically died on air while presenting his breakfast show today.

Tim Gough, 55, was an hour into the program when the music stopped half way through a song.

Lubbock Avalanche-Journal

Lubbock native, prima ballerina NaTalia Johnson dies at 37

LeAnda Staebner   Lubbock Avalanche-Journal
Published 6:15 p.m. CT May 20, 2021 | Updated 6:19 p.m. CT May 20, 2021

≡   billboard   🔍

Colombian Singer Andres Cuervo Dies

10/11/2022

According to his agency, the 40-year-old singer died of a heart attack in his apartment in Paris, where he had been residing. "

THE HIU

Young Canadian reporter dies suddenly and unexpectedly of 'heart failure'

Mike Fahey   9/24/2022

The Telegraph

Queen's guard found dead in barracks at 18 years old

30 September 2022 • 9:44pm

NEW YORK POST   LOG IN

ABC News' 'This Week' producer Dax Tejera dead at 37

By Patrick Reilly   December 24, 2022 | 7:40pm | Updated

NEWS SAN DIEGO

ABC 10News producer Erica Gonzalez passed away overnight

Dec 21, 2022

VARIETY

Jan 5, 2022 4:15am PT

Kim Mi-soo, Korean Actor in 'Snowdrop,' Dies at 29

People   SUBSCRIBE

Celebrity Trainer Eric Fleishman Dead

November 28, 2022 10:13 PM

TMZ

'FLASHDANCE,' 'FAME' SINGER IRENE CARA
DEAD

11/26/2022 5:00 AM PT

RollingStone   SUBSCRIBE

Jo Mersa Marley, Grandson of Bob Marley and Son of Stephen, Dead at 31

BY JON BLISTEIN
DECEMBER 27, 2022

ECHO

Young actress Emily Chesterton dies suddenly

08:01, 11 NOV 2022   UPDATED 09:16, 11 NOV 2022

TMZ

'DOG'S MOST WANTED' STAR
DAVID ROBINSON DIES DURING ZOOM CALL

12/1/2022 / 10 AM PT

NEW YORK POST   LOG IN

TikTok star Megha Thakur dead at 21: 'Beautiful inside and out'

By Jack Hobbs   December 1, 2022 | 12:58pm | Updated

DEADLINE   TIP

Actress Yakira Chambers Collapses and Dies Suddenly

December 5, 2022 4:48pm

"

Since I got my vaccine, I have a problem, I have a series of problems. As a result, I can't train, I can't play. So now I regret having taken the vaccine, but I couldn't have known.

— Champion tennis star, Jeremy Chardy

"

'Please pray for me': Justin Bieber reveals he's been struck by facial paralysis from rare syndrome and shares fears as he struggles to eat after being forced to cancel tour dates

By JUSTIN ENRIQUEZ FOR DAILYMAIL.COM
PUBLISHED: 15:27 EDT, 10 June 2022 | UPDATED: 10:59 EDT, 11 June 2022

☰ **GLAMOUR** 🔍

Hailey Bieber Was Hospitalized for a Blood Clot in Her Brain

March 12, 2022

CONFIRM THIS STORY IS REAL

CONFIRM THIS STORY IS REAL

Note: Similar reactions associated with COVID vaccines have been reported to CDC and FDA by the thousands, including hundreds specifically about Ramsay Hunt Syndrome.

Note: Similar reactions associated with COVID vaccines have been reported to CDC and FDA by the thousands

DIED ON STAGE

SUR in English

World of magic in mourning after Arsenio Puro collapsed on Madrid stage and died

Monday, 17 October 2022, 13:18

CONFIRM THIS STORY IS REAL

NEW YORK POST   LOG I

Singer Mikaben dead at 41 after collapsing on stage during a show

By Nika Shakhnazarova   October 18, 2022 | 3:28am | Updated

CONFIRM THIS STORY IS REAL

ABC
Culture

04/10/2022

Bin Valencia, legend of Argentine rock, dies during performance with his wife and children

Another similar tragedy has occurred the same day in Seville, where the singer of Antiheroes Rock Band has died after suffering a heart attack on stage

CONFIRM THIS STORY IS REAL

A rock singer dies after suffering a heart attack during concert in Seville

By taketonews

OCT 2, 2022

CONFIRM THIS STORY IS REAL

THE DAILY WIRE

– NEWS –

Talented High School Singer Collapses And Dies On Stage While Performing At All-State Event

By Ryan Saavedra · Oct 19, 2022 DailyWire.com

Daniel Moshi was performing Friday at the All-State Honors Show Choir for the Illinois American Choral Directors Association at Naperville North High School when he suddenly collapsed.

CONFIRM THIS STORY IS REAL

NEW YORK POST   LOG IN

Valencia Prime, 25, collapses and dies while onstage

By Dana Kennedy   September 17, 2022 | 5:27pm | Updated

CONFIRM THIS STORY IS REAL

ET

British Actress Josephine Melville Dies Backstage After Performing in Play

By **Samantha Schnurr** 8:35 AM PDT, October 24, 2022

News

Conductor collapses and dies mid-performance at leading German opera house

25 July 2022, 11:15 | Updated: 25 July 2022, 12:08

Jammu: Artiste collapses and dies on stage during performance

In the video, Yogesh, dressed as a woman, is performing an energetic number on stage. After a few moments, he collapsed on the floor

FP Trending | September 08, 2022 19:12:59 IST

News in Germany

Lifestyle

Melanie Müller: Ballermann singer had a stroke

1 month ago

Roberto Gonzalo (Holy Land) suffers a heart attack and collapses on stage at his concert in Zaragoza

3 October, 2022 10:56 am

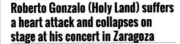

NEWS   SPORTS   HIP HOP   WATCH   PHOTOS   TOURS

BRANDY

HOSPITALIZED FOR POSSIBLE SEIZURE

≡ **Pitchfork**

NEWS

The Avalanches Cancel 2022 Tour Due to "Serious Illness"

October 1, 2022

≡   **NME**   Q

John Paul Young cancels tour dates due to "health issues"

The 'Love Is in the Air' singer has "decided to take a back seat for a time" after recently being admitted to hospital

By **Alex Gallagher** | 30th August 2022

VARIETY

Log in ▼

HOME > MUSIC > NEWS   Aug 23, 2022 8:08am PT

Former Cradle of Filth Guitarist Stuart Anstis Dies at 48

By Jem Aswad ∨

BBC 👤   Home   News   Sport   Reel   ···   Q

Limp Bizkit postpone European tour over Fred Durst health concern

18 July

☰   **Ripples**Nigeria   Q

ENTERTAINMENT

'Gangster Paradise' rapper, Coolio dies of heart attack

Published 4 weeks ago on September 29, 2022
By **Adekunle Fajana**

TODAY   ● TODAY all day   Q   ☰

POP CULTURE

Comedian Heather McDonald suffers skull fracture after collapse during show

THE SOCIETY OF ACTUARIES RESEARCH INSTITUTE:
Another Independent Source Confirms Our Thesis

On August 17, 2022 the Society of Actuaries Research Institute (SOA) published their Group Life COVID 19 Mortality Survey Report.

This report represents approximately 80% of the group life US revenues. In Table 5.7 of their report, you can see there was an event that occurred in the Third Quarter of 2021 (highlighted in red and orange) which confirmed excess mortality of 78% and 100% respectively, for two age groups (25-34 and 35-44). In March we saw 84% excess mortality into Fall 2021, and though we already knew our analysis was solid, here it was verified by SOA. It is undeniably clear that an event occurred in the Third Quarter of 2021, the same period that vaccine mandates were ordered by the Biden Administration, and enforced by corporate America.

Table 5.7
EXCESS MORTALITY BY DETAILED AGE BAND

Age	Q2 2020	Q3 2020	Q4 2020	Q1 2021	Q2 2021	Q3 2021	Q4 2021	Q1 2022	4/20-3/22	% COVID	% Non-COVID	% Count
0-24	116%	124%	104%	101%	119%	127%	110%	91%	111%	3.3%	8.1%	2%
25-34	127%	132%	121%	118%	131%	178%	131%	125%	133%	13.3%	19.6%	2%
35-44	123%	134%	128%	129%	133%	200%	156%	136%	142%	23.1%	19.2%	4%
45-54	123%	127%	129%	133%	119%	180%	151%	143%	138%	27.4%	10.8%	9%
55-64	117%	123%	130%	130%	114%	153%	141%	137%	131%	24.0%	6.7%	18%
65-74	117%	115%	133%	130%	108%	131%	125%	122%	122%	18.6%	3.9%	17%
75-84	114%	114%	133%	123%	106%	119%	121%	121%	119%	14.0%	4.6%	20%
85+	112%	103%	124%	111%	92%	104%	105%	103%	107%	10.3%	-3.5%	27%
All[11]	116%	115%	129%	123%	107%	134%	126%	122%	121%	17.1%	4.3%	100%

Note that our early analysis was based upon the CDC's All-Cause Mortality Data, whereas the SOA data is from death claims in the Group Life Divisions of insurance companies that were surveyed. This is important because they are two different databases and two different population sets which both show the same excess death rates among millennials (ages 25-44).

Some weeks later, SOA published another report that again makes my case in the clearest terms:

> *"One of the features of 2021 was that excess mortality was worse for the working ages. We see that the impact on the U.S. population between 15 and 64 was 31.7% worse..."*

The obvious explanation for the deaths moving to younger, healthier people is that most of those who were employed and insured were forced by mandate to take an experimental vaccine product to maintain their employment, even if they were hesitant, or had a medical or religious objection. And those who were unemployed, self-employed or retired had a choice!

Note: Readers interested in a much deeper dive into the shift of deaths from older to younger, including a full analysis of the extensive underlying data, are directed to Appendix Two, Page 182.

GLOBAL COVID DEATHS:
Another Independent Source Confirms Our Thesis

The Johns Hopkins Coronavirus Resource Center (CRC) and the Johns Hopkins Center for Systems Science and Engineering (CSSE) track and analyze COVID data worldwide, coordinating with researchers and faculty from schools of Medicine, Public Health, Nursing, and Engineering.

The chart you'll see next represents CSSE's assessment of daily COVID deaths worldwide, from the beginning of the pandemic up to the point that the 20 largest countries had all administered one or two doses (March 2022).

Dear Public Health Officials: With 68% of the world's population vaccinated and 13-billion doses administered, if the vaccines are safe and effective, how do you explain that the overwhelming number of COVID deaths –and all the highest peaks in deaths– occurred after commencement of mass vaccination?

Daily New Confirmed COVID-19 Deaths per Million People

7-day rolling average. Due to varying protocols and challenges in the attribution of the cause of death, the number of confirmed deaths may not accurately represent the true number of deaths caused by COVID-19.

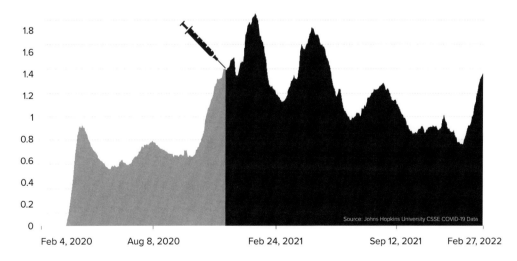

Source: Johns Hopkins University CSSE COVID-19 Data

This next chart shows the progress of mass vaccination in the world's 20 most populous countries up to March 2022. Countries above the 100-mark had administered more than one dose to some or all of their citizens. Countries above the 200-mark had administered more than two doses.

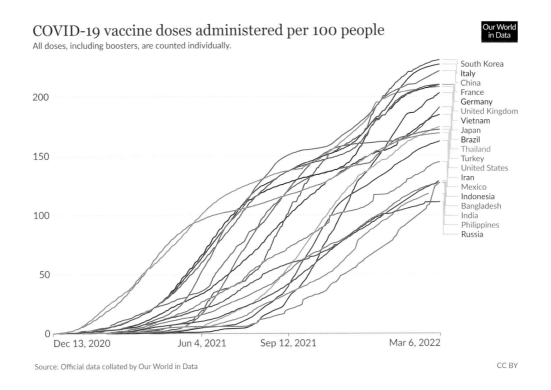

COVID-19 vaccine doses administered per 100 people
All doses, including boosters, are counted individually.

Source: Official data collated by Our World in Data   CC BY

BRITAIN'S OFFICE OF NATIONAL STATISTICS:
Another Independent Source Confirms Our Thesis

As more media outlets reported my thesis that the new vaccine products were causing a spike in excess deaths, impressive experts from around the world volunteered to test my conclusions. What you sometimes find in finance are people willing to give a clear-eyed look at reality, including unpopular reality that others aren't yet willing to consider.

Among the key experts who joined our analysis group is Carlos Alegria, a physicist who turned to quantitative finance (and has a PhD in both). Well known in the European hedge fund community, Carlos was recruited for stints with the prestigious firms Winton Capital Management and Mansard Capital.

Carlos wrote the book *Economic Cycles, Debt, Demographics: The Underlying Macroeconomic Forces that will Shape the Coming Decades*. As the title implies, he is a master at making sense of large datasets, and also a person willing to look around the corner no matter what's waiting there. That was about to be tested.

There is no evidence right now that healthy children or healthy adolescents need boosters – no evidence at all.

— Chief Scientist World Health Organization,
Soumya Swaminathan

Offit told me that getting boosted would not be worth the risk for the average healthy 17-year-old boy. Offit advised his own son, who is in his 20s, against getting a third dose.

— The Atlantic Magazine Interview
Dr. Paul Offit, longtime FDA advisor

FOR CHILDREN:
A SMOKING GUN

As we turned our attention to children, another PhD physicist joined us: Yuri Nunes. I first met Carlos and Yuri when we began to explore forming the hedge fund company, Phinance Technologies (See Appendix Three, Page 188). They immediately offered to help assist the research I was already doing on excess All-Cause Mortality. Carlos and Yuri set out to study the excess death data from the UK's Office of National Statistics (ONS) to detect any illuminating trends in deaths during 2020, 2021 and 2022.

When they analyzed the stats for children ages 1-14, their findings were stunning.

In addition to breaking down the data by age group, the physicists obtained vaccine uptake data. The next chart is 2020 excess deaths in the UK for ages 1 to 65. The first 12 weeks of the year represent the pre- pandemic period; not surprisingly, it shows zero excess deaths in all age groups.

Weeks 13-17 show the effect of the first COVID peak (original variant) in March-April, and the effect of the first pandemic lockdowns undertaken to "flatten the curve." We see a rise in excess deaths in some age groups but a decline in excess deaths for children 1-14 years old. When excess deaths in the post-lockdown summer period stabilize for the older age groups, they continue to trend downward for children. Stay with me.

When excess deaths for older people rise in October, November, December, deaths among children continue to decline, which is easily explained by two solidly established facts: First, COVID does not affect this age group from a mortality standpoint, so deaths of children had no reason to go up. And second, since the most common cause of death in this age group has always been accidents, the fact of lockdowns, school closures, remote schooling and a general curtailment of activity meant fewer opportunities for children to die in accidents. The next chart shows deaths of children continuing to decline.

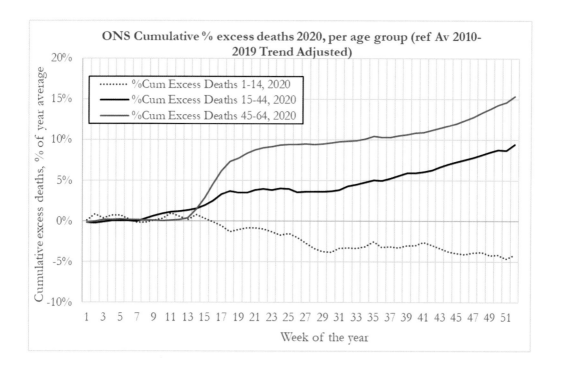

The next chart is excess UK deaths for ages 1-14, 15-44 and 45-65 in 2021 – and here we see something startling: While the 2020 chart above reflects the period of excess deaths attributed to COVID (where we saw declining excess deaths among children), the 2021 chart includes the period after vaccination.

The arrows illustrate when mass vaccination commenced for each age-group. Remember, mass vaccination was phased by age, starting around week 13 for the 45-64 age group, week 18 for the 15-44 age group, and week 33 for children, starting with the 12-15 age cohort.

Notice that after vaccination of the age groups 15-44 and 45-65, the slope (rate of change) of both lines begins to rise. That's quite odd given that the Delta and Omicron strains are less virulent than the earlier strains, and given the widely promoted idea that the vaccines were "safe and effective." If safe and effective were the reality, then the slopes of the older lines should be falling and heading towards zero – not heading to new highs.

Notice how excess deaths continue to trend downward even after pandemic restrictions were lifted in Spring and Summer 2021. Next, notice with the introduction of the first dose of COVID vaccines for age group 12-15, there is *a trending rise in excess deaths*. If you have a supportable theory other than the commencement of vaccination, a theory to explain that tragic trend, please let the world know about it.

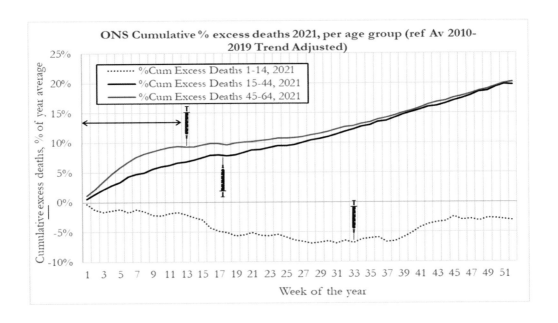

The most stunning thing in the next chart is that mortality continues to rise throughout the year for all age groups during the least virulent variant (Omicron), during what should have been the end of the COVID Pandemic.

Deductive reasoning calls out to us clearly: The sad trend is associated with the vaccination, especially for children.

No credible person would try to argue that Omicron, the mildest strain, had suddenly decided to target children. Those who want to argue an increase in accidental deaths among children will find the actual culprits are labeled "Other" and "Cause Unknown."

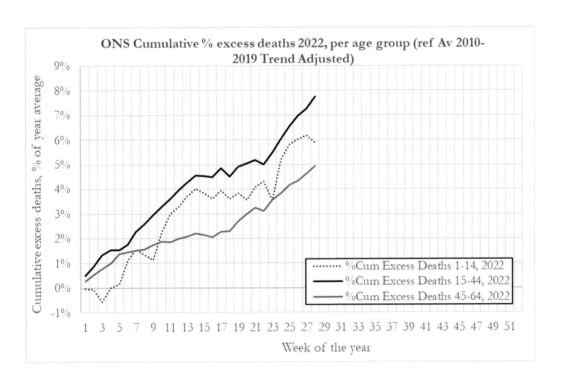

ONS Cumulative % excess deaths 2022, per age group (ref Av 2010-2019 Trend Adjusted)

.... %Cum Excess Deaths 1-14, 2022
——— %Cum Excess Deaths 15-44, 2022
——— %Cum Excess Deaths 45-64, 2022

Cumulative excess deaths, % of year average

Week of the year

We were not the only people on Earth to observe these alarming and tragic trends. On September 8, 2022, **the UK Government stopped offering COVID vaccines to children under 12.**

Also, the UK Government warned two years ago against use of COVID vaccines for pregnant women:

> "Use in women of childbearing potential could be supported provided healthcare professionals are advised to **rule out known or suspected pregnancy prior to vaccination**. Women who are breastfeeding should also not be vaccinated."

And yet, at the time of this writing, the United States (now followed by the UK, despite their previous positions) promotes, encourages, and mandates these experimental vaccines for pregnant women, and for people in all age groups – literally down to six months old. From the CDC website at the time of printing:

> "COVID-19 vaccination is recommended for all people 6 months and older. This includes people who are pregnant, breastfeeding, trying to get pregnant now, or might become pregnant in the future."

And CDC throws some fear into the mix as well, warning that if you're pregnant and catch COVID (which COVID vaccines won't prevent anyway), "...you are at increased risk of complications that can affect your pregnancy and developing baby." They don't bother to mention anything about increased risk of stillbirth and miscarriage associated with the COVID vaccines, and in case you don't remember this from back on page 56:

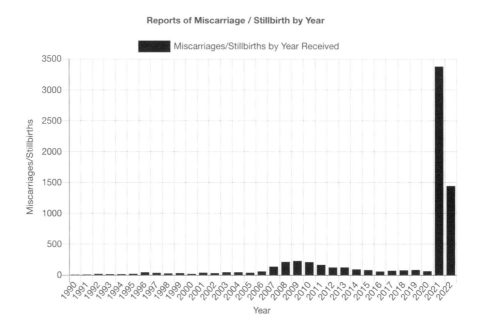

DENMARK STOPS COVID VACCINE FOR THOSE UNDER 50:
Another Independent Source Confirms Our Thesis

A month after the UK stopped offering COVID vaccines to children under 12, Denmark's health officials went even further and ended mass vaccination for everyone under 50. (They had long before stopped recommending the vaccine for people under 18.)

From the Danish Government website:

> *"The purpose of vaccination is not to prevent infection with COVID-19, and people under 50 are therefore currently not being offered booster vaccination."*

> *"In addition, younger people aged under 50 are well protected against becoming severely ill from COVID-19, as a very large number of them have already been vaccinated and have previously been infected with COVID-19, and there is consequently good immunity among this group."*

Several interesting points to note: First, the public was initially told —stridently and repeatedly— that the vaccines prevented infection and transmission. Then Denmark makes a statement that would have seemed ridiculous if said about any previous vaccine: "The purpose of vaccination is not to prevent infection with COVID-19," admitting in Orwellian doublespeak that what they promoted about the vaccines had been false. (Same for many other governments, including the US.)

The newer untruth they promote is that the vaccine prevents severe illness and death, an unprovable myth also pushed by the US Government and Big Pharma. But there is no data to credibly support the claim, as all the fully vaccinated people whose deaths are attributed to COVID would tell you – if they could.

The second interesting point in the Denmark statement is implied: They are now saying it is better for young people to get infected with COVID than take the experimental shots. The Danish health authorities clearly see something about the vaccines that they are not overtly sharing with the public.

Indeed, Danish politicians and officials have plenty they'd want to keep out of public awareness. After making Denmark one of the world's most vaccinated countries, the data is grim:

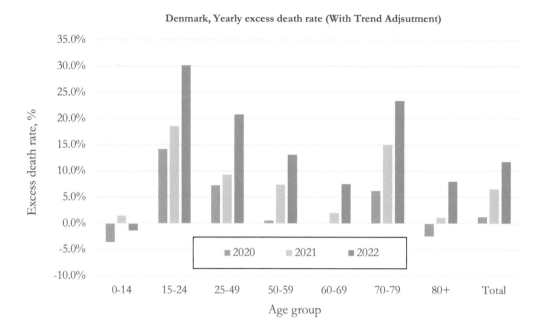

As you're seeing, excess deaths are on the rise after mass vaccination, for all ages. If vaccines were safe and effective, why would Denmark see its trend in excess deaths get (way) worse, and at just the time COVID variants became less deadly?

Notice also that in 2022, young people 15-24 have the highest death rate of any group. It can't be that COVID has suddenly become deadly to teenagers, because if that were case, why ever would the Danish Government have just stopped vaccination for the young?

The Danish authorities are stopping the vaccine for people under 50 because they see this same alarming data you're seeing in these pages. Undeniably, something new is causing excess deaths amongst the young. Though admitting mass vaccination is the culprit would be very difficult for any government, it would be the truth.

On Wall Street, we'd call Denmark's recent decisions a *tell*, or a *tone change*. During my career, whenever I interviewed CEOs or studied corporate press releases, they never came out and said, "Hey guys, our growth is slowing and next year's earnings are going to drop." Instead, they hope and pray nobody notices as they sell their own stock before the unwelcome news hits.

So, when we see a huge tone change by the Danish Government, what they are actually communicating is that the vaccine doesn't work, kills some people – and they know it.

Physicists Carlos Alegria and Yuri Nunes did an excellent and cogent analysis of Denmark's death rate trends. Speaking generally, death rates in every developed country had been coming down in the past decade, likely due to advances in technology, medicine, and safety. But in both of the next charts, the favorable historical trend is broken, with the blue lines showing a new and unfavorable trend.

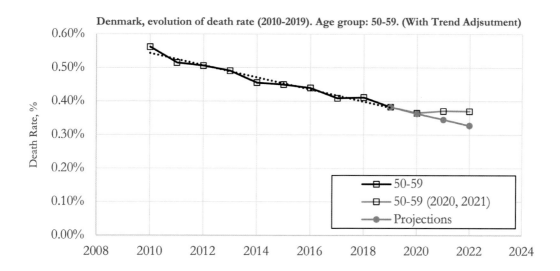

How does Denmark look from a total population standpoint? Sadly, it looks like an epic train wreck: Excess death has gone up each year since mass vaccination began. In these graphs, you see the Danish healthcare system going backwards – and you don't need any government authority to explain this data to you.

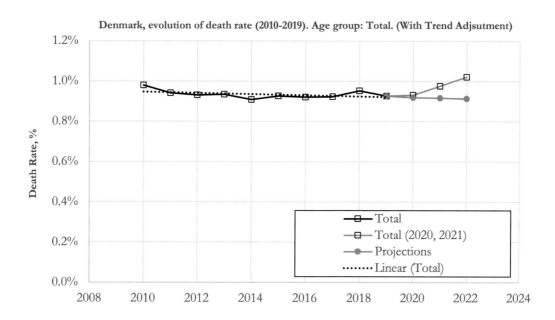

NEWS YOU LIKELY MISSED...

THE WALL STREET JOURNAL.

Are Covid Vaccines Riskier Than Advertised?

There are concerning trends on blood clots and low platelets, not that the authorities will tell you.

By Joseph A. Ladapo and Harvey A. Risch
June 22, 2021 1:09 pm ET

One remarkable aspect of the Covid-19 pandemic has been how often unpopular scientific ideas, from the lab-leak theory to the efficacy of masks, were initially dismissed, even ridiculed, only to resurface later in mainstream thinking. Differences of opinion have sometimes been rooted in disagreement over the underlying science. But the more common motivation has been political.

CONFIRM THIS STORY IS REAL

Covid Vaccines Up to 100 Times More Likely to Cause Serious Injury to a Young Adult Than Prevent It, Say Top Scientists

BY WILL JONES 7 SEPTEMBER 2022 7:00 AM   SHARE

CONFIRM THIS STORY IS REAL

Newsweek   SUBSCRIBE FOR $1 >

U.S.

More Study Urged After 7 Teen Boys Suffer Heart Inflammation After Getting COVID-19 Vaccine

BY MARY ELLEN CAGNASSOLA ON 6/4/21 AT 1:20 PM EDT

U.S.   PFIZER   CORONAVIRUS   TEENAGERS

CONFIRM THIS STORY IS REAL

Daily Mail.com   health

Up to one in 7,000 American teens suffered heart inflammation after their Covid vaccine, study suggests

- Scientists at Kaiser Permanente reviewed 340 cases out of 6.9 million vaccinees
- Found those getting a second dose were most likely to suffer myocarditis
- This struck within the first seven days in the vast majority of cases

By LUKE ANDREWS HEALTH REPORTER FOR DAILYMAIL.COM
PUBLISHED: 17:02 EDT, 3 October 2022 | UPDATED: 17:11 EDT, 3 October 2022

Thousands of American teenagers may have suffered heart inflammation after getting a Covid jab, a study suggests.

Researchers found up to one in 7,000 boys aged 12 to 15 years old developed myocarditis after receiving the Pfizer vaccine.

CONFIRM THIS STORY IS REAL

FOX NEWS

FOX NEWS FLASH · Published July 22, 2022 7:29pm EDT

Dr. Deborah Birx says she 'knew' COVID vaccines would not 'protect against infection'

The former White House COVID response coordinator downplays vaccine efficacy

CONFIRM THIS STORY IS REAL

Months Before Vaccine Rollout, UK Government Knew There Would be Many Adverse Reactions

In October 2020, the Medicines and Healthcare Products Regulatory Agency (MHRA) issued a tender for "*an Artificial Intelligence software tool to process the expected high volume of Covid-19 vaccine Adverse Drug Reactions (ADRs)*."

Despite knowing the vaccines would cause serious harm, the UK Government went ahead promoting and requiring the vaccines, and passed them off as "Safe and Effective."

CONFIRM THIS STORY IS REAL

The government-funded British Broadcasting Corporation established and runs this very active coalition of major companies that have "*pledged to work together to **tackle harmful misinformation about Covid-19 vaccines**.*"

A key member of TNI is James Smith, former CEO of the Reuters news agency, current Chairman of the Reuters Foundation that oversees **Reuters Fact Check**, and... current member of **Pfizer's Board of Directors**.

News stories that call attention to unexplained sudden deaths among young vaccinated people are clearly bad for Pfizer, and bad for those encouraging mass vaccination. TNI companies work hard to eliminate and limit such stories, while promoting information from government agencies and Pharma.

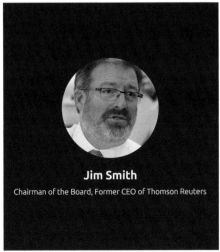

Jim Smith
Chairman of the Board, Former CEO of Thomson Reuters

Newsweek

NEWS

State Surgeon General Warns Young Men COVID Vaccines Pose 'High Risk' of Death

BY **FATMA KHALED** ON 10/8/22 AT 5:46 PM EDT

Florida's Surgeon General Joseph A. Ladapo warned on Friday against young men receiving COVID-19 vaccines, citing a disputed analysis by the state health department that they pose an "abnormally high risk" of death.

CONFIRM THIS STORY IS REAL

Autopsy Confirms NY State College Student Died From "COVID-19 Vaccine-Related Myocarditis"

(TJV NEWS) 24-year-old New York college student George Watts Jr. died on October 27, 2021, due to complications related to the Pfizer Covid-19 shots he took in August and September.

It was revealed recently that the Bradford County Coroner's Office listed the COVID vaccine as the cause of death.

CONFIRM THIS STORY IS REAL

The Atlantic

SCIENCE

DID A FAMOUS DOCTOR'S COVID SHOT MAKE HIS CANCER WORSE?

A lifelong promoter of vaccines suspects he might be the rare, unfortunate exception.

By Roxanne Khamsi

SEPTEMBER 24, 2022

CONFIRM THIS STORY IS REAL

 REUTERS®

4 minute read · November 18, 2021 1:31 PM PST · Last Updated a year ago

Wait what? FDA wants 55 years to process FOIA request over vaccine data

(Reuters) - Freedom of Information Act requests are rarely speedy, but when a group of scientists asked the federal government to share the data it relied upon in licensing Pfizer's COVID-19 vaccine, the response went beyond typical bureaucratic foot-dragging.

As in 55 years beyond.

CONFIRM THIS STORY IS REAL

Daily **Mail**.com

Revealed: PR firm that represents Pfizer and Moderna also sits on CDC vaccine division - sparking major conflict of interest concerns

PUBLISHED: 12:16 EDT, 12 October 2022 | UPDATED: 14:27 EDT, 12 October 2022

CONFIRM THIS STORY IS REAL

False Statistical Modeling
Used to Change Our World

2001: Neil Ferguson, a modeler at Imperial College London predicted **150,000** people would die from foot and mouth disease

Actual number of deaths: **200**

2002: Neil Ferguson predicted up to **156,000** deaths in the UK from Mad Cow Disease

Actual number of deaths: **177**

2005: Neil Ferguson predicted that up to **200 million** would die from bird flu

Actual number of deaths: **282** (over 6 years)

2009: Neil Ferguson predicted that swine flu would kill **65,000** in the UK

Actual number of deaths: **45**

2020: Neil Ferguson predicted up to **179,000** COVID deaths in Taiwan in the first full year of pandemic

Actual number of deaths: **10**

Despite decades of dramatic and persistent failures, Neil Ferguson's prediction that as many as 2-million Americans would die from COVID in 2020 was used to justify lockdowns, school closures, social distancing, and all that followed.

(These same failed models are used to support claims that many millions of lives have been saved by COVID vaccines.)

Pilot of Boeing flight to St. Petersburg dies suddenly on board plane

By Matthew Roscoe · 19 September 2022 · 16:08

CONFIRM THIS STORY IS REAL

Pilot Dies Suddenly After Takeoff from Chicago Airport

 By Richard Moorhead

November 27, 2022 at 8:41am

An American Airlines flight was forced to return to its take-off location Nov. 26 after the aircraft's pilot suffered a medical emergency in-flight.

CONFIRM THIS STORY IS REAL

NEW STRAITS TIMES

Helicopter pilot dies after losing consciousness during rescue mission

By Zahratulhayat Mat Arif - September 11, 2022 @ 8:30pm

CONFIRM THIS STORY IS REAL

Simple Flying

Home > Airline News > Biman Bangladesh Pilot Dies After Inflight Medical En

Pilot Dies After Inflight Medical Emergency

BY JAKE HARDIMAN   PUBLISHED SEP 5, 2021

Photo: Md Shafuzzaman Ayon via Wikimedia Commons

CONFIRM THIS STORY IS REAL

Moderna COVID-19 Vaccine Effectiveness Turns Negative Within Months: Study

October 10, 2022   Updated: October 11, 2022

The effectiveness of Moderna's COVID-19 vaccine against infection turns negative over time, according to a new study that was funded by the vaccine maker.

CONFIRM THIS STUDY IS REAL

Pfizer Exec Concedes COVID-19 Vaccine Was Not Tested on Preventing Transmission Before Release

By Jack Phillips | October 11, 2022 Updated: October 13, 2022

A A̤   Print

A Pfizer executive said Oct. 10 that neither she nor other Pfizer officials knew whether its COVID-19 vaccine would stop transmission before entering the market last year.

CONFIRM THIS STORY IS REAL

ScienceDirect

Volume 40, Issue 40, 22 September 2022, Pages 5798-5805

Serious adverse events of special interest following mRNA COVID-19 vaccination in randomized trials in adults

3.4. Harm-benefit considerations

In the Moderna trial, the excess risk of serious AESIs (15.1 per 10,000 participants) was higher than the risk reduction for COVID-19 hospitalization relative to the placebo group (6.4 per 10,000 participants). [3] In the Pfizer trial, the excess risk of serious AESIs (10.1 per 10,000) was higher than the risk reduction for COVID-19 hospitalization relative to the placebo group (2.3 per 10,000 participants).

CONFIRM THIS STUDY IS REAL

Actuaries raise alarm that Australians are unexpectedly dying at incredibly high rate

BY RHODA WILSON ON DECEMBER 16, 2022 • (67 COMMENTS)

The Australian government should be urgently investigating the "incredibly high" 13% excess death rate in 2022, the country's peak actuarial body says.

CONFIRM THIS STORY IS REAL

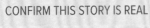

Why are so many footballers collapsing? There has been a worrying spike in cardiac arrests and stars retiring with heart-related issues, but leading sports cardiologist insists it is NOT to do with Covid vaccine

PUBLISHED: 17:34 EDT, 15 December 2021 | UPDATED: 19:38 EDT, 15 December 2021

Nine years separated Marc-Vivien Foe's death from a heart attack on a pitch in Lyon to Fabrice Muamba's near-fatal collapse at Tottenham, and another nine passed before Christian Eriksen was brought back to life at Euro 2020 this summer.

Three of the most harrowing in-game episodes that football has seen were spread over more than 18 years, yet it feels that barely a week goes by at the moment without news of another cardiac-related incident in the game.

CONFIRM THIS STORY IS REAL

 GlobalResearch
globalresearch.ca / globalresearch.org

Massachusetts Death Certificates Show Excess Mortality Could be Linked to COVID Vaccines

By Madhava Setty

Global Research, November 23, 2022

CONFIRM THIS STORY IS REAL

LIFESTYLE.INQ

Data suggests sudden adult death syndrome due to vaccines

By: Rafael Castillo-M.D.   Philippine Daily Inquirer / 07:21 PM June 27, 2022

CONFIRM THIS STORY IS REAL

≡ Watch Live   Latest Video

High school cancels rest of football season due to lack of healthy players

Published: Oct. 13, 2022 at 6:03 PM PDT |

On Tuesday, Bellevue High School Athletic Director Jim Hicks announced that the school made the decision because the team has a limited number of healthy players available.

CONFIRM THIS STORY IS REAL

 politics

President Joe Biden tests positive for Covid-19 again

By Kevin Liptak and Donald Judd, CNN
Updated 3:59 PM EDT, Sat July 30, 2022

JUST IN

CONFIRM THIS STORY IS REAL

 NEW YORK POST LOG IN

CDC director Dr. Rochelle Walensky comes down with COVID a month after booster shot

By Mary Kay Linge
October 22, 2022 | 1:33pm | Updated

The director of the Centers for Disease Control and Prevention has tested positive for COVID-19 — one month after she publicly celebrated getting her booster shot.

CONFIRM THIS STORY IS REAL

---

yahoo!finance

Pfizer CEO Contracts COVID-19 Infection For Second Time

Vandana Singh
September 26, 2022 · 1 min read

- **Pfizer Inc's** (NYSE: PFE) CEO Albert Bourla said he had tested positive for COVID-19.

- Bourla has received four doses of the COVID vaccine developed by Pfizer and **BioNTech SE** (NASDAQ: BNTX).

CONFIRM THIS STORY IS REAL

---

  politics

First on CNN: Jill Biden tests positive for rebound case of Covid-19

By Kate Bennett, CNN
Updated 9:59 PM EDT, Wed August 24, 2022

Biden, who is currently in Rehoboth Beach, Delaware, tested positive again for Covid-19 with an antigen test on Wednesday. She tested negative during a routine test on Tuesday.

CONFIRM THIS STORY IS REAL

WJ THE WESTERN JOURN

News

Triple-Vaxxed Justin Trudeau Tests Positive for COVID-19 Again

By Jack Davis• June 13, 2022 at 4:55pm

The positive test was Trudeau's second bout with COVID-19 this year. He tested positive in January, which was also the month when he received his third dose of the coronavirus vaccine.

CONFIRM THIS STORY IS REAL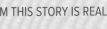

Townhall

Fully Vaccinated and Double-Boosted Dr. Fauci Has COVID

 Spencer Brown
June 15, 2022 3:45 PM

Despite the fact that President Biden previously stated that COVID-19 is "a pandemic of the unvaccinated," Dr. Fauci is just one of many within the Biden administration who have contracted COVID after being fully vaccinated.

CONFIRM THIS STORY IS REAL

Report reveals Pfizer shot caused avalanche of miscarriages, stillborn babies

Among the first reports handed over by Pfizer was a 'Cumulative Analysis of Post-authorization Adverse Event Reports' describing events reported to Pfizer up until February 2021.

Pfizer's report states that there were 23 spontaneous abortions (miscarriages), two premature births with neonatal death, two spontaneous abortions with intrauterine death, one spontaneous abortion with neonatal death, and one pregnancy with "normal outcome." That means that of 32 pregnancies with known outcome, 28 resulted in fetal death.

CONFIRM THIS STORY IS REAL

Babies Dying In Scotland: Govt Issues Investigation Into Rising Infant Mortality Rate

Written by Kashish Sharma | Updated : October 4, 2022 10:49 AM IST

The Scottish government has issued an investigation into the suddenly rising infant mortality rate in the country. There have been two spikes over a period of six months. Earlier the figures had shown that the death rate for babies under one year of age was at its highest in 10 years. The increase was high enough to initiate an investigation.

CONFIRM THIS STORY IS REAL

People

LIFESTYLE > HEALTH

New Study Shows COVID-19 Vaccine Does Cause Changes to People's Menstrual Cycles

The new study confirms findings that there is a link between vaccination against COVID-19 and an average increase in menstrual cycle length

By Amanda Taylor Published on September 28, 2022 09:13 PM

CONFIRM THIS STORY IS REAL

 ISRAEL NATIONAL NEWS
ARUTZ SHEVA + A

New study: COVID vaccination shown to decrease sperm counts

Effect begins 2 months following vaccination & persists for at least 5 months, when study ended.

Y Rabinovitz

Jun 22, 2022, 3:51 PM (GMT+3) 7INN
ARUTZ SHEVA

vaccine   vaccination   Coronavirus

A new study conducted by Israeli researchers at Shamir Medical Center, Tel Aviv University, Herzliya Medical Center, and Sheba Medical Center, has published disturbing findings regarding the possible impact of COVID vaccination on male fertility.

CONFIRM THIS STORY IS REAL

CALGARY | News

Deaths with unknown causes now Alberta's top killer: province

Published July 5, 2022 6:16 p.m. ET
Updated July 6, 2022 8:40 p.m. ET

 By Nicole Di Donato
CTV News Calgary Multimedia Journalist

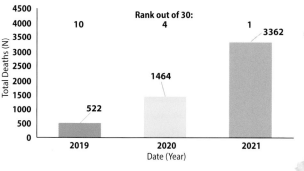

Cause of death: "Other ill-defined and unknown causes of mortality"

Rank out of 30:

CONFIRM THIS STORY IS REAL

THE SCIENCE TIMES

HOME > MEDICINE & HEALTH

30 Deaths Among Nursing Home Residents in Norway: Here's What Possibly Caused Such Occurrences

Olive Marie Jan 19, 2021 01:03 AM EST

CONFIRM THIS STORY IS REAL

DAILY NEWS

ONLY $1 FOR 6 MONTHS
In-depth election news

LOG IN

CORONAVIRUS

Nearly two dozen nursing home residents in Norway die within days of getting COVID vaccine

By Jessica Schladebeck
New York Daily News
Jan 15, 2021 at 1:53 pm

CONFIRM THIS STORY IS REAL

the bmj

Covid-19: Pfizer-BioNTech vaccine is "likely" responsible for deaths of some elderly patients, Norwegian review finds

BMJ 2021 ; 373 doi: https://doi.org/10.1136/bmj.n1372
(Published 27 May 2021)
Cite this as: *BMJ* 2021;373:n1372

CONFIRM THIS STORY IS REAL

"There is a genuine risk that the time of death will be brought forward and the patient will suffer adverse reactions to the vaccine. It may often be better not to vaccinate."

— Norway Government Report on Nursing Home Residents

While the Norway Government urged caution, Canada (and the US) required COVID vaccines for all nursing home patients. In Canada, nearly 70% of the people whose deaths were attributed to Covid were residents of nursing homes.

CBC | MENU ˅

NEWS

Ottawa

Canada's nursing homes have worst record for COVID-19 deaths

Posted: Mar 30, 2021 1:00 AM PDT | Last Updated: March 30, 2021

CONFIRM THIS STORY IS REAL

THE US Sun

SICKLY JAB Doctor dies and 36 others 'develop rare blood disorder after getting Moderna and Pfizer Covid vaccines'

0:06 ET, Feb 10 2021 | Updated: 10:54 ET, Aug 27 2021

The New York Times

Johns Hopkins Scientist: 'A Medical Certainty' Pfizer Vaccine Caused Death of Florida Doctor

Dr. Jerry L. Spivak, an expert on blood disorders at Johns Hopkins University, told the New York Times Tuesday that he believes "it is a medical certainty" that Pfizer's COVID vaccine caused the death of Dr. Gregory Michael.

According to the New York Times:

"Dr. Jerry L. Spivak, an expert on blood disorders at Johns Hopkins University, who was not involved in Dr. Michael's care, said that based on Ms. Neckelmann's description, 'I think it is a medical certainty that the vaccine was related.'

"We don't believe at this time that there is any direct connection to the vaccine."

— Pfizer

---

 msn

 Washington Examiner   + Follow   View Profile

NIH scientists received estimated $350 million in royalties since 2009: Report

Story by Jenny Goldsberry • May 10   👍   👎   💬 Comments

The institute and roughly 1,800 of its scientists have received an estimated $350-$400 million in these payments, from entities like pharmaceutical companies, during the last decade.

CONFIRM THIS STORY IS REAL

≡   **The New York Times**

F.D.A.'s Drug Industry Fees Fuel Concerns Over Influence

The pharmaceutical industry finances about 75 percent of the agency's drug division, through a controversial program that Congress must reauthorize by the end of this month.

Sept. 15, 2022

CONFIRM THIS STORY IS REAL

Thai princess in line to throne has devastating heart attack while out jogging – 44 years old

04:08, 15 Dec 2022 | UPDATED 13:31, 15 Dec 2022

CONFIRM THIS STORY IS REAL

CNN health

FDA vaccine advisers 'disappointed' and 'angry' that early data about new Covid-19 booster shot wasn't presented for review last year

Updated 9:24 AM EST, Wed January 11, 2023

"I was angry to find out that there was data that was relevant to our decision that we didn't get to see," said Dr. Paul Offit, a member of the Vaccines and Related Biological Products Advisory Committee, a group of external advisers that helps the FDA make vaccine decisions. "Decisions that are made for the public have to be made based on all available information – not just some information, but all information."

Science

SCIENCEINSIDER | PLANTS & ANIMALS

FDA no longer requires animal testing before human drug trials

10 JAN 2023

New medicines need not be tested in animals to receive U.S. Food and Drug Administration (FDA) approval,

---

5000 Students Screened with EKG: 17% Had at Least One Cardiac Symptom After 2nd Dose

Deadline Passes For Pfizer To Submit Results Of Post-Vaccination Heart Inflammation Study To US Regulators

THURSDAY, JAN 12, 2023 - 06:44 AM

Daily Mail health
Home | U.K. | News | Sports | U.S. Showbiz | Australia | Femail | Health | Science | Money | Video | Travel | Shop

Moderna begins trialing mRNA shot that is injected directly into the HEART to treat heart attack patients

12 January 2023

---

*"Those of you who think the vaccine kills people can **use me as a test. If I die, you were right**."*

— Fitness Expert Doug Brignole

Bodybuilding icon Doug Brignola Found Dead at Home

October 14, 2022

"This week has been a difficult one in the Massachusetts Law Enforcement community.

"Three active-duty Law Enforcement Officers all passed away suddenly within the last four days."

 -Massachusetts State Police

WESTERN Standard

Three Massachusetts law enforcement officers die suddenly within four days

Jan 6, 2023

MRNA Vaccines Increase Risk of Contracting COVID-19; Each Booster Shot Raises Risk Even More in Study of 51,000 Cleveland Clinic Workers

A study published Monday at medRxiv shows that MRNA vaccines raise the risk of contracting COVID-19 and that each MRNA vaccine booster increases the risk of contracting COVID-19, while those who have not received any MRNA vaccine have the lowest risk of contracting COVID-19.

December 21, 2022

CNN entertainment   Movies   Television   Celebrity

Jamie Foxx is hospitalized after 'medical complication'

April 13, 2023

Daily Sun

Bronny James, son of LeBron James, suffers cardiac arrest at USC basketball practice

Jul 25, 2023

USA TODAY

Alabama high school basketball star Caleb White dies after collapsing during pickup game

Aug. 12, 2023

Sun

SUDDEN DEATH Instagram influencer Larissa Borges, 33, dies suddenly after suffering 'double cardiac arrest' that left her in coma

Aug 30 2023

NEW YORK POST

Olympic swimmer Helen Smart was found dead by 4-year-old daughter: 'Distraught'

August 17, 2023

RollingStone

Tori Kelly Rushed to Hospital Due to Blood Clots

JULY 25, 2023

USA TODAY

South African rapper Costa Titch dies hours after collapsing during festival performance

March 12, 2023

NEWS

Jansen Panettiere, actor and younger brother of Hayden Panettiere, died from an enlarged heart, his family says

Feb 27, 2023

NEW YORK POST

TikTok star Jehane Thomas dies suddenly at 30, suffered migraines for months

March 20, 2023 | 10:13am

People

Former Illinois News Anchor, 42, Dies After Sudden Illness on Vacation

Published on April 14, 2023 11:31 AM

Ex-beauty queen, 25, who collapsed and died at Michael Owen's stables

14:19, 29 Mar 2023

Mirror

Daily Mail .com

Callie Mitchell dies a week after being found unconscious from heart condition at cheer camp in Texas

11 August 2023

CALGARY HERALD

'Eric was loved': Family, teammates mourn death of 13-year-old athlete

Published Nov 15, 2022

NEWS

20-year-old New Mexico State soccer player found dead at her home

July 13, 2023

NEW YORK POST

High school football player taken off life support after collapse

July 8, 2023

Daily Mail.com

World's sixth best young water-skier 'dies suddenly' aged just 18

PUBLISHED: 18:19 EDT, 18 May 2023

NEW YORK POST

Surprising new details about Ryan Mallett's death as harrowing body cam video released

June 29, 2023 | 12:47am | Updated

Mail Online

Two-time national champion boxer Jude Moore dies suddenly aged just 19

PUBLISHED: 06:14 EDT, 21 March 2023

FOX

Local footy match abandoned after 26yo player taken to hospital due to mid-game cardiac arrest

April 24th, 2023 9:40 am

Daily Mail

Oklahoma boy aged just 14 suffers a STROKE while attending wrestling camp leaving him on ventilator after he underwent emergency brain surgery

PUBLISHED: 17:40 EDT, 28 June 2023

FOX NEWS

Dominican basketball player who previously blamed COVID vaccine for rare heart condition dies of heart attack

Published June 24, 2023 7:32pm EDT

REGINA LEADER-POST

Regina's sports community mourns sudden passing of Theo Gibbs

Published Jan 06, 2023

People

Runner Steve Shanks Dies 'Out of the Blue' Hours After Completing London Marathon

Published on April 26, 2023 04:13 PM

NEW YORK POST

UNLV football player Ryan Keeler dead at 20

February 21, 2023

Paul Offit, longtime FDA advisor:

There certainly is a causal link between vaccination and myocarditis and pericarditis. No doubt about it. It's unclear why... it may be, as was actually noticed in 2020, that SARS-CoV-2 virus spike protein mimics one of the proteins on heart muscle cells, specifically the heavy chain of actin. So if that's true, then while you're making an immune response to the SARS-CoV-2 spike protein, you're also inadvertently making an immune response to your own heart muscle.

BBC
Former NFL player Chris Smith dies at 31

🕐 18 April

cyclingnews
Heart irregularities force Jan Polanc to retire from professional cycling

May 18, 2023

Mail Online
Ex-New York Mets minor league pitcher Matt Pobereyko found dead at 31 after suffering a heart attack

February 2023

BBC
Liverpool: Footballer hit by cardiac arrest saved by opposition player

🕐 4 May

Devastated family pay tribute to son, 15, who died during football match

15 Mar 2023

Mirror

WSAZ 3 NewsChannel
High school senior dies after suffering heart attack at school, officials say

Jan. 10, 2023

The Scarborough
Former Whitby man and leading surgeon dies suddenly aged 46

Published 12th Jan 2023, 10:24 BST

INDEPENDENT
An Alabama college student died at an apartment in the US Virgin Islands'

Tue, July 25, 2023

Des Moines Register
19-year-old Iowa soldier dies in Georgia after medical emergency during US Army basic training

Published 7:23 a.m. CT July 17, 2023

Sun

HOLIDAY TRAGEDY Brit student Rhea Hourigan, 19, dies from cardiac arrest

Published: 9:56 ET, May 5 2023

N•Q Report

National Guard Soldier Suffers TWO Heart Attacks After Moderna "Vaccine"

April 9, 2023

NEW YORK POST

Conn. teachers thought 5-year-old boy who collapsed during recess was just 'playing dead': lawsuit

April 9, 2023 | 12:08pm

abcNEWS

Teen goes into cardiac arrest at cheer competition and her mom saves her life

March 16, 2023, 2:55 PM

Family heartbreak as daughter suddenly collapses and dies in dad's arms with 'heart attack'

1 Mar 2023

Daily Mail

Mourners gather for the funeral of Thai cave survivor who died aged 17 after being found in his dormitory

4 March 2023

REUTERS

Healthcare & Pharmaceuticals

CDC reports fewer COVID-19 pediatric deaths after data correction

March 18, 2022

Children accounted for about 19% of all COVID-19 cases, but less than 0.26% of cases resulted in death, according to the American Academy of Pediatrics, which summarizes state-based data.

Business Standard

Peru declares 90-day health emergency over Guillain-Barre Syndrome outbreak

Last Updated : Jul 11 2023 | 4:38 AM IST

PMC PubMed Central®

A case of fatal multi-organ inflammation following COVID-19 vaccination

Published online 2023 Mar 20.

/ CITY JOURNAL

May 09 2023

The Harm Caused by Masks

A new study suggests that the excess carbon dioxide breathed in by mask-wearers can have major health consequences.

Evidence continues to mount that mask mandates were perhaps the worst public-health intervention in modern American history. While concluding that wearing masks "probably makes little or no difference" in preventing the spread of viruses, a recent Cochrane review also emphasized that "more attention should be paid to describing and quantifying the harms" that may come from wearing masks. A new study from Germany does just that, and it suggests that the excess carbon dioxide breathed in by mask-wearers may have substantial ill-effects on their health—and, in the case of pregnant women, their unborn children's.

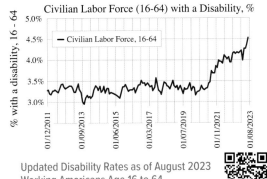

Civilian Labor Force (16-64) with a Disability, %

Updated Disability Rates as of August 2023
Working Americans Age 16 to 64

May 2022

Global excess deaths associated with COVID-19, January 2020 - December 2021

A comprehensive view of global deaths directly and indirectly associated with the COVID-19 pandemic.

NIH National Library of Medicine
National Center for Biotechnology Information

PubMed®

A prospective study on myocardial injury after BNT162b2 mRNA COVID-19 fourth dose vaccination in healthy persons

European Journal of
Heart Failure

HFA
Heart Failure
Association

Research Article   🔒 Open Access (cc) (i) (=) (S)

Sex-specific differences in myocardial injury incidence after COVID-19 mRNA-1273 booster vaccination

Natacha Buergin, Pedro Lopez-Ayala, Julia R. Hirsiger, Philip Mueller, Daniela Median, Noemi Glarner, Klara Rumora, Timon Herrmann, Luca Koechlin, Philip Haaf, Katharina Rentsch, Manuel Battegay, Florian Banderet, Christoph T. Berger, Christian Mueller ✉

First published: 20 July 2023
https://doi.org/10.1002/ejhf.2978

 Urban Care

Switzerland No Longer Recommends COVID-19 Vaccination: Here's Why

FDA Detects Serious Safety Signal for COVID-19 Vaccination Among Children

May 25, 2023

Heart Scarring Observed in Children Months After COVID-19 Vaccination: Study

 Zachary Stieber, Reporter
Aug 5 2023

TORONTO SUN

Canadian teen dies suddenly in Mexico

Published Feb 23, 2023

ENGLISH Jagran

Two Indian Pilots Die In A Day, One Collapsed Moments Before Flying, Another Mid-Air

The DGCA is investigating both incidents to determine the cause of the cardiac arrests.

Thu, 17 Aug 2023

INDEPENDENT

Pilot dies in bathroom on Miami flight carrying 271 passengers

August 17, 2023

Daily **Mail**.com

Veteran British Airways pilot dies after suffering heart attack in hotel shortly before he was due to captain flight from Cairo to Heathrow

PUBLISHED: 06:31 EDT, 12 March 2023

The Dallas Morning News

Southwest Airlines pilot requires medical attention during Las Vegas to Columbus flight

9:18 AM on Mar 23, 2023 CDT

USA TODAY

Pilot of wayward plane that crashed in Virginia was slumped over, fighter jet pilots said

Jun 5

People

11-Year-Old Dies After Losing Consciousness on Turkish Airlines Flight Headed to New York

Published on June 12, 2023 11:09AM EDT

VOK

Virgin Australia Pilot Taken Ill Just 30 Minutes After Takeoff, Prompting Emergency Landing

7TH MARCH 2023

nzherald.co.nz

Air Albania flight attendant, 24, dies suddenly after plane lands

24 Feb, 2023 04:39 PM | 3 mins to read

Anthony Fauci on New Vaccines, 1999

If you take it and then a year goes by and everybody's fine, then you say, 'Okay, that's good, now let's give it to five hundred people.' And then a year goes by and everything's fine, so you say, 'Well then now let's give it to thousands of people.'

And then you find out that it takes 12 years for all hell to break loose.

*And then **what have you done?***

THE EUROPEAN STATISTICAL OFFICE:
Another Independent Source Confirms Our Thesis

Denmark provided us with a valuable case study, but it's hardly the only European country with an excess death problem. Recent analysis of Austria, Spain, Italy, Netherlands, Sweden, Belgium and France all reveal their excess deaths occurring in the same two phases:

1. The COVID Pandemic
2. Mass vaccination

After mass vaccination, death rates of working-age people increased in every one of these countries. I challenge anyone to find a virologist who'd suggest that a virus would switch to targeting the younger healthier population in the second year of a pandemic. (Clearly, the Danish Government didn't think so when they ended mass vaccination for everyone under 50.)

Looking at this next chart, it is undeniable that something happened to increase deaths in 2021 and 2022, something new, something that wasn't an influencing factor in 2020.

Excess Deaths (%) During Pandemic from 2020 to 2022
Before Vax / After Vax, 25-49 Years of Age

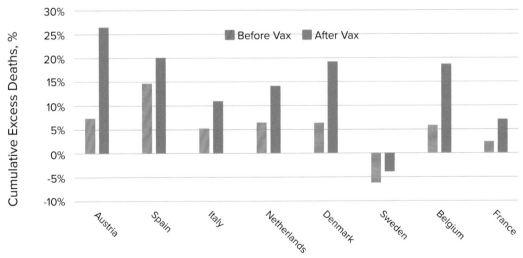

As people search to explain the spike in deaths, particularly people unwilling to imagine or even contemplate that mass administration of a brand new pharmaceutical product could be the culprit, the typical explanations proposed include drug overdoses, suicides, delayed diagnosis of other diseases, and missed screenings of rapidly-fatal cancers. These ideas, whether alone or in combination, have not been borne out to explain the sad excess mortality. In any event, these ideas cannot explain the glaring temporal reality: In Summer and Fall 2021, using America as an example, **there was an 84% spike in excess deaths among the youngest working people (25 - 44), and that spike immediately followed the massive vaccine mandate.**

For deaths to spike in any or all of the proposed categories within that exact same time period for any reason other than mass vaccination is statistically impossible. We know that hundreds of millions of doses were administered, and there is no other factor that affected nearly all working-age people simultaneously.

Next, again using the US as an example, the SOA report shows that the healthiest population (working age Americans) experienced an excess death rate 8% <u>higher</u> than the general US population. And note that the general population is far less healthy than working Americans. Since the intellectually lazy reasons proposed (suicide, delayed medical care, etc) cannot be responsible for these eye-popping increases all happening at the same time, there is an urgent question being ignored every day by public health officials: What is causing this new and tragic loss of life among young working-age people?

TWO PANDEMICS
Year 1: COVID
Year 2: Vaccines

Recall when President Biden proclaimed "a pandemic of the unvaccinated." Turns out he was exactly, precisely, literally 100% wrong. Politicians love catchphrases; in this case the accurate one would be "Pandemic of the Vaccinated."

Europe and America currently have a second pandemic far worse than the first, because its numbers are worse, and also because its victims are healthy young people who would not otherwise have died.

The next chart shows excess death rates for young working people in eight European countries, focusing first on the period during which mass vaccination commenced, and then the same 3-month period a year later, after millions more were vaccinated and boosted.

The year-over-year comparisons show that all these countries experienced more than their usual death rates after mass vaccination commenced. And seven of the eight countries experienced substantial jumps in excess deaths; as vaccination continued and increased, their rate of excess deaths also trended upward.

Given that COVID is now less virulent than it was at the start of vaccination, all countries' death rates should be trending back towards zero excess death. Also, vaccination rates are now much higher than in 2021, so if "safe and effective" is true, death rates should be plummeting, not increasing. But they continue to rise. Again (and again), logic and deductive reasoning indicate that mass vaccination is the often-deadly culprit for some otherwise healthy people in the prime of their lives.

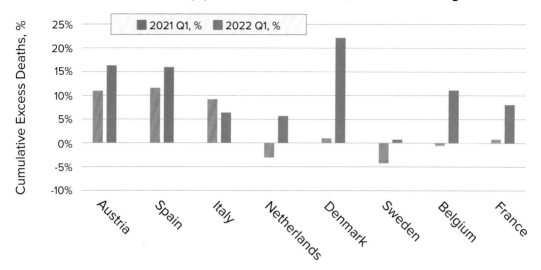

Excess Deaths (%) in Q1 2021 vs Q1 2022, 25-49 Years of Age

Note: For an excellent in-depth report on how Excess Death rates are computed:

DISABILITY

HIGHEST-EVER DISABILITY
RATES IN AMERICA

Esteemed expert Dr. Jessica Rose has done extensive analysis of the US Government's Vaccine Adverse Event Reporting System (VAERS). Commenting on the devastating disabilities she has seen, Dr. Rose has said, "Although death is the most extreme adverse event, there are some things that might actually be worse than death." Before I dive into the disability data, let's look with compassion at those who have suffered from serious adverse events that didn't end in death.

DR.DREW

Pro Mountain Bike Racer Discusses Severe COVID-19 Vaccine Reaction

December 8, 2021

CONFIRM THIS STORY IS REAL

7NEWS.com.au

Gold Coast soccer community rallies around girl, 14, after suffering heart attack on field

Warren Barnsley / Soccer / Updated 17.10.2021

CONFIRM THIS STORY IS REAL

Sport•Chlen.fr

A young player suffered a heart attack in the middle of a match

Published on5 Sep 21 at 16:17

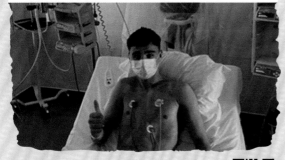

CONFIRM THIS STORY IS REAL

CYCLINGTIPS

SONNY COLBRELLI HOSPITALIZED AFTER CARDIAC ARREST

MARCH 21, 2022

CONFIRM THIS STORY IS REAL

INDEPENDENT

Emil Palsson: Footballer collapses from cardiac arrest during game in Norway

Tuesday 02 November 2021 05:41

CONFIRM THIS STORY IS REAL

CBS | NCAA BB

USC five-star freshman Vince Iwuchukwu sidelined indefinitely after suffering cardiac arrest this summer

Sep 29, 2022 at 2:50 pm ET • 2 min read

CONFIRM THIS STORY IS REAL

Aargauer Zeitung

Cardiac arrest in the substitute goalie: SC Zofingen game was not kicked off

René Achterberg suffered a cardiac arrest during the warm-up before the second division home game against FC Grenchen on Saturday and was able to be resuscitated.

4/10/2022, 5:46 p.m.

CONFIRM THIS STORY IS REAL

  ouest france   Search city, news, misc

The 13-year-old teenager has a cardiac arrest while playing football in Rennes: two passers-by intervene

It is a very rare accident which took place on Wednesday March 30, 2022, at the end of the day, at the Parc de Maurepas, in Rennes. A 13-year-old football player suffered a heart attack. Two general practitioners who were passing by immediately performed cardiac massage while waiting for help

CONFIRM THIS STORY IS REAL

EUROSPORT

SWISS ATHLETE IRENE CADURISCH COLLAPSES DURING WOMEN'S BIATHLON RELAY

PUBLISHED 16/02/2022 AT 06:28 GMT-7

CONFIRM THIS STORY IS REAL

nope   ORF.at

Player collapsed

At a football match, player Milos Jovanovic suddenly collapsed and swallowed his tongue. The club's masseur reacted immediately and saved his life.

April 16, 2022 1:35 p.m. (Update: April 16, 2022 4:01 p.m.)

CONFIRM THIS STORY IS REAL

sn.dk

Vordingborg player received heart massage on the pitch

25-year-old Benjamin Rud Jensen collapsed in the grass and is now admitted to the cardiac ward.

November 7, 2021, 8:54 p.m | Updated November 8, 2021, 11:02 am

CONFIRM THIS STORY IS REAL

≡ VeloNews

Audrey Cordon-Ragot pulls out of UCI Road World Championships after suffering stroke

SEPTEMBER 17, 2022

CONFIRM THIS STORY IS REAL

EuroWeekly NEWS
NEWS AND VIEWS SINCE 1998

Swiss Olympic athlete Fabienne Schlumpf diagnosed with myocarditis

By Matthew Roscoe · 11 January 2022 · 10:18

EuroWeekly NEWS
NEWS AND VIEWS SINCE 1998

Swiss athlete Sarah Atcho diagnosed with heart problems

By Matthew Roscoe · 19 January 2022 · 12:12

The Guardian

Wigan's Charlie Wyke thanks manager for helping to save his life after collapse

- CPR from Leam Richardson and club doctor kept striker alive
- Wyke out of hospital after procedure following cardiac arrest

Thu 2 Dec 2021 14.53 EST

FOX 8

Coaches perform CPR, use defibrillator on High Point student who collapsed during basketball practice

by: Tess Barsebuhr
Posted: Jan 31, 2022 / 06:06 PM EST
Updated: Feb 1, 2022 / 02:45 PM EST

"As a coach, I've never experienced in 20 years a medica magnitude," said Scott Van Newkirk, a science teacher a at the school.

The Athletic

World No. 2 women's golfer Nelly Korda diagnosed with blood clot

March 13, 2022 Updated 12:55 PM PDT

VERVE TIMES

Queensland Boy, 15, Suffers Heart Attack On Soccer Field, Parents Perform CPR

By Albert Wagner · On Nov 24, 2021

Norra Skåne

Wrestling star Lukas Ahlgren: "I thought I was going to die"

Wrestling/Kristianstad · Published May 24, 2022

Daily Record

Young rugby player thanks team-mates and medics for saving his life

Hamish Bell (20) nearly died after he suffered a cardiac arrest during a training session

11:02, 20 JUL 2021

FOX 2 DETROIT

Detroit police officers reunite with runner who suffered cardiac arrest they saved

By David Komer online producer | Published November 4, 2021 | Updated 11:17PM

KLEINE ZEITUNG

After the collapse, Florian Ploner was allowed to leave the hospital

The 22-year-old's symptoms indicated a heart attack or stroke.

September 05, 2021 at 4:20 p.m

ABC News Journalist Reveals She Developed Heart Condition Due to Covid Vaccine

Published December 22, 2022 at 1:45pm

The Sydney Morning Herald

Dr Kerryn Phelps says she suffered COVID vaccine injury, calls for more research

Former federal MP Dr Kerryn Phelps says she and her partner experienced vaccine injury.

December 20, 2022

Daily Mail .com

7 December 2022

Rod Stewart reveals his son Aiden, 11, was rushed to hospital in an ambulance with a suspected heart attack after collapse at football match

English cricket star, 18, survives mystery heart attack on beach

14:20, 15 Nov 2022

Mirror

MUNDODEPORTIVO

12/13/2022

Marta Collado, successfully operated on for tachycardia

ESPN

SCORES

Tennessee's Tamari Key out for season with blood clots in lungs

Dec 8, 2022

LIFESTYLE MAGAZINE

Famous Influencer Eliezer discovered heart disease after experiencing severe chest pain

BASKET EUROPE

French international Ana Tadic's career in limbo due to heart problem

The CDC director last year said if we vaccinate a million children, there might be 30 or 40 cases of mild myocarditis. And they said, if you get myocarditis from COVID, that's worse or happens at a higher rate. But that's not true. The studies have come out. Europe reacted by banning the Moderna vaccine altogether in young people, in many parts of Europe and everybody under 30. We're now learning that there's significant heart damage, 31% of people having physical activity restrictions. 63% of children after myocarditis had evidence of heart swelling months down the road on MRI. So we were playing with fire.

— Dr Marty Makary, Johns Hopkins University

US BUREAU OF LABOR STATISTICS:

Another Independent Source Confirms Our Thesis

We all know intuitively that healthy working people don't suddenly experience life-altering physical disabilities without accidents or injury involved. The kinds of stories you've just seen were not part of my young adulthood experience, and it's easy to discern that something changed in 2021. That was dramatically affirmed (I should say reaffirmed) when Josh Stirling and I discovered an important database collected by the US Bureau of Labor Statistics.

Every month, the Department of Labor conducts thousands of household surveys asking questions that relate to disability. This is a real-time survey in which respondents report if they themselves are disabled or if someone in the household is disabled. Importantly, these statistics are independent of things like medical claims or doctors' notes, thus giving an exceptionally clear and up-to-the-moment picture of the disability trend in the US. Also important: The data is independent of the Pharma-captured public health establishment.

Prior to mass vaccination, the run rate was about 29-30 million disabled Americans of working age, give or take, for the last five years. Since mass vaccination commenced in 2021, the number of disabled Americans has increased to almost 33 million. That means nearly 3.5 million more Americans of working age are now too disabled to work.

Several of the charts that follow come from a study done by Carlos and Yuri at the end of July 2022.

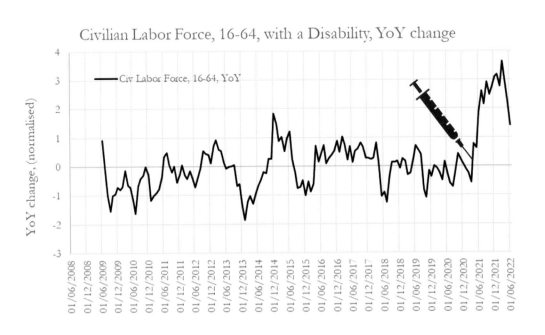

That chart paints a picture worth a thousand words. One sees that the year-over-year rate of change oscillated around a steady average going back to 2008, and then accelerates dramatically in May of 2021. For math geeks, that's a stunning 3 standard deviation event.

On the next chart (published by another independent source, the Federal Reserve Bank), anyone can see the sharp rise in disabilities well above the 5-year average.

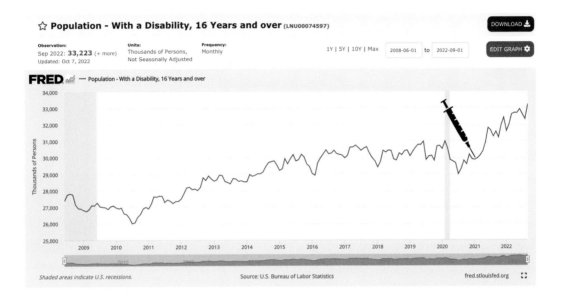

The trend on the next chart is equally grim: You see that the current disability rate has not only increased 6.6% over the historical average disability rate, but it's reached a higher level than at any time in at least 15 years.

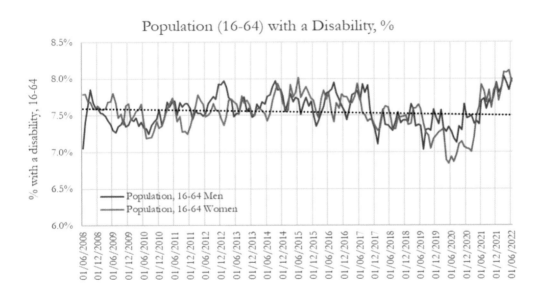

The next chart shows the employed population by different ages. Focus on the black line, which reflects ages 16-64, a cohort we know (used) to be healthier than the overall US population.

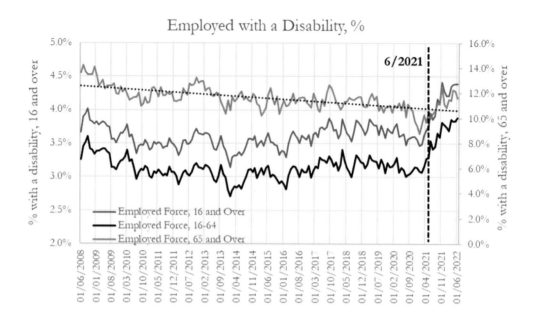

The next chart shows that working women in America are faring much worse than men (13% worse than men).

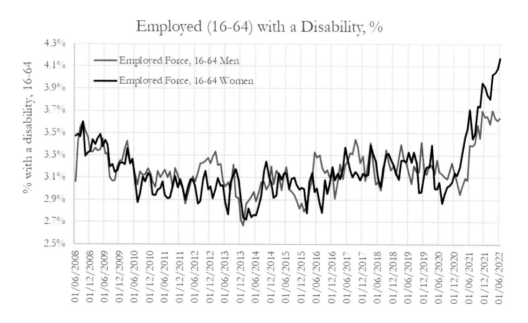

The bottom line revealed in the charts you've just seen is that the healthier employed folks experienced a greater increase in rate of disability (22.6% increase) than the comparable total US population (6.6% increase). Obvious question of the day: Why are healthier folks seeing a bigger increase in the rate of their disabilities starting around May of 2021? What else was increasing at a substantial rate in mid-2021? Mass vaccination. If you have credible data that points to something else that will explain all of what you've seen here, please let the world know about it.

In the meantime, the disability crisis has knock on effects throughout our economy – and everyone who has seen Help Wanted signs at businesses knows this first-hand. You can't simply eliminate millions of workers without profound consequences.

This is a short book precisely because it's tightly focused on just a few important topics. I have provided several vantage points from which to look at the undeniable increase in deaths among the young. I avoided reliance upon any one source, or any one idea. Rather, I checked indicators and trends of different types, from various sources, and one after another, source after source, the results of analysis are complimentary, meaning they point in the same direction.

Any reader can confirm that 1) Healthy young people have been dying and becoming disabled with alarming frequency; and 2) The rate of these tragedies is new and unusual, and not sufficiently explained by government officials; and 3) Glaring public health questions are not being asked or answered by those in power.

Curious readers who are satisfied that these three conclusions are factual reality will naturally want to know **why** events have unfolded as they have. I've carefully avoided expressing opinions and theories, and have explored these disturbing topics by the numbers, literally. Of course, we all have opinions and theories about our new world, and I'm no different. But: I made a conscious decision to stay away from theorizing about **why** and **who**. Being loyal to that commitment, there are just a few observations I'd like to express:

> At this point, public health officials and vaccine makers are aware of everything that you now know from this book – at a minimum. At this point, they have moved beyond something we could write off as incompetence, in that they are allowing (and even forcing) mass use of products they know are harmful. At this point, powerful people in public health and Pharma are in full cover-up mode. They almost have to be, because it's hard to imagine how they could pivot from what they've done to what they ought to do.

At this point, the negligence is criminal.

Ed Dowd

AFTERWORD
by Gavin de Becker
Bestselling Author, *The Gift of Fear*

A quick thought experiment:

Imagine that thousands of healthy young Americans died suddenly, unexpectedly, mysteriously – and then kept dying at an alarming and escalating rate. (Once upon a time), that would trigger an urgent CDC inquiry to determine the cause of the deaths. Imagine attentive and curious public health officials discover the decedents had all repeatedly ingested a new and little-understood drug. Next, the officials determine to a certainty that the drug these kids took has a clear mechanism of action for causing inflammation of the heart and other cardiac injuries in some people. They learn that public health officials in other countries have seen the same thing and stopped recommending this same drug to young people. Next, some of the most senior and revered scientific advisors to the US Government publicly recommend the drug be stopped for young people. Finally, thousands of doctors around the world sign petitions and write op-eds opposing the drug for young people. Experts from Harvard, Yale, MIT, Stanford, and Oxford universities come forward to voice their concerns.

Alas, that thought experiment doesn't require any imagination, because it's exactly what's occurred – except for the part about attentive and curious CDC officials rushing in to inquire. That part I had to make up.

In the pre-COVID world, wouldn't inquisitive reporters chase such a story, and wouldn't the FDA pause administration of the new mystery drug until a comprehensive inquiry was complete? And above all, **wouldn't such a drug have quickly become a leading suspect worth considering for its possible role in the deaths**?

Somehow, those have become rhetorical questions.

But not to Edward Dowd. His pioneering exploration of these sad sudden deaths was months ahead of the Wall Street Journal story about a lethal conundrum facing insurance companies. Apparently, as mass vaccination progressed in 2021, excess death claims in working-age Americans tripled. Given the temporal relationship, might there be a connection between vaccines and these deaths? Apparently not, because the WSJ story didn't even mention mass vaccination among the causes worth considering:

1. Delayed medical treatment from 2020
2. People's fear of seeking treatment
3. Trouble lining up appointments
4. Drug abuse and other societal troubles

5. People not taking care of themselves
6. Long COVID
7. Not-yet-known long-term effects of COVID
8. People dying later "from the toll COVID has taken on their bodies"

Numbers 1, 2, and 3 are all subsets of the same concept: impact of lockdowns and fear. Numbers 6, 7, and 8 are all subsets of the same concept: the impact of COVID illness.

Aside from drug abuse and trouble getting doctors' appointments, did anything else happen in 2021 that might possibly-maybe-perhaps-call-me-crazy be worth considering?

The insurers attributed most of the 2021 excess deaths to heart and circulatory issues, neurological disorders, and stroke. It's a coincidence, apparently, that nearly all of their deceased customers had just been injected with something known to cause heart and circulatory issues, neurological disorders, and stroke. Head-scratching by insurance industry CEOs and experts interviewed for the WSJ article was understandable, because, you know, how in the world could deaths possibly be linked to some brand new, never-before-used drug that was minimally tested, maximally rushed, mass-administered, and oh yeah, and by the way, is known to cause the very medical issues their customers were dying of?

(If you have any doubt as to whether the mRNA vaccines cause cardiac problems, see Appendix Four, Page 190, for a sampling of 100 published papers on vaccine-induced cardiac injuries to young people.)

Please forgive three more rhetorical questions:

- Don't Americans have plenty of excellent reasons to distrust Big Pharma companies?
- Why is Pharma being afforded the kind of trust previously reserved for companies that <u>don't</u> have a long history of criminal fraud?
- What's up with that?

Some relevant recent history: In 2021, a jury convicted Elizabeth Holmes of medical fraud. During the trial, she admitted that her company conducted medical tests using the same old machines that were supposedly being replaced by her remarkable new technology.

Though people were given fraudulent and incorrect medical results, the news media's fawning over Holmes and Theranos kept the fraud alive for years, such that the company was eventually valued at $9 billion. Not a typo.

Theranos would not have been possible if the news media were skeptical or even curious when it comes to medicine and medical technology. As you scan these few examples of news media promotion, consider that Vioxx and opioids and COVID vaccines were all similarly floated into our society on a soft cushion of media praise, free of scrutiny or skepticism.

<u>USA Today:</u>

Elizabeth Holmes is tall, smart and single. Well, maybe not truly single. "I guess you should say I'm married to Theranos," Holmes says with a laugh. Only she's not kidding... while Holmes is a billionaire on paper, nothing seems to interest her less... "We're successful if person by person we help make a difference in their lives," says Holmes, who has a soft yet commanding voice that makes a listener lean in as if waiting for marching orders.

New Yorker:

Although she can quote Jane Austen by heart, she no longer devotes time to novels or friends, doesn't date, doesn't own a television, and hasn't taken a vacation in ten years… "I have done something, and we have done something, that has changed people's lives... I would much rather live a life of purpose than one in which I might have other things but not that."

CNN:

The company she founded has the potential to change health care for millions of Americans.

Forbes:

Elizabeth Holmes, 30, is the youngest woman to become a self-made billionaire – and she's done so four times over. "What we're about is the belief that access to affordable and real-time health information is a basic human right, and it's a civil right," she says.

Since the news media couldn't hardly change course after all that worship, it wasn't any journalist who undid the Theranos scam. It was Stanford professor John Ioannidis who was willing to publicly point out that Theranos hadn't published any peer-reviewed research about their products.

(To bring us to the present moment for a moment, that same John Ioannidis, esteemed physician, scientist and epidemiologist, was among the first and most vocal critics of lockdown policies. For that, he's been the target of excoriation and cancellation by news media and the medical/Pharma cabal. *Don't you dare wake us from our COVID fever dream, Professor!*)

Back to Theranos: The company attempted to boost its credibility by getting then-Vice President Biden to visit their facility. In order to conceal the lab's true operating conditions, Holmes and her team created a fake lab for the Vice President to tour.

The Theranos case should remind us that 25% of drugs approved by the FDA are later pulled from the market – so they can't all be miracle drugs. Still, whatever a public health bureaucrat says today, the news media repeats, defends, and then enshrines as fact. Never has their lack of skepticism and curiosity about their biggest sponsors been on such colorful display as during the last two years, during which they've constantly promoted and parroted false Pharma claims related to new vaccines, false claims about the new Pfizer COVID treatment, false claims about Ivermectin, and every other false claim by Pharma and government. Real journalism being MIA has left us helplessly living in a world of compliance rather than science.

Here's a fast journey through some intentionally forgotten history: After the FDA approved Vioxx, there were many litigations related to the inconvenient fact that the drug doubled the risk of heart attack. How many litigations? Oh... 27,000 of them, but who's counting? Risk of heart attack sounds familiar, given that hundreds of studies have now found mRNA vaccines increase the risk of cardiac death in young males. But again, who's counting?

Merck was eventually forced to withdraw Vioxx, and was ordered to pay criminal fines of almost a billion dollars for overstating the drug's safety with a now familiar refrain: "safe and effective."

Like today, when CDC and FDA are receiving hundreds of thousands of reports of adverse reactions to COVID vaccines (e.g. myocarditis, stroke, blood clots, death), Merck told jury after jury that heart attack deaths had nothing whatsoever to do with their wonder-drug. They fought lawsuits like... well, like a Pharma company, accusing plaintiffs of falsifying data. Pot/kettle.

When a jury awarded one widow $253 million, Merck appealed, and that award was overturned. A bunch of other lawsuits followed, with Merck winning some, losing some – until a class-action lawsuit concluded that Merck had violated the law by selling a drug that was unfit for sale, because of, you know, doubling the risk of heart attack, or some such thing. And then...

Merck agreed to a mass tort settlement of ***$4.85 billion*** to end thousands of individual lawsuits. And then...

Merck announced a settlement with the US Attorney's Office, resolving the $950 million fine that had been levied against the company.

Did that end it? Nope, litigation with seven states remains outstanding today. But the real punchline is...

Vioxx is returning to market. Yes. We can look forward to breathless news media reports about another (likely renamed) wonder-drug that's safe and effective.

The story of Vioxx is not a cautionary tale; rather, it's a regular event. A group of concerned scientists wrote about it in 2017:

> *To increase the likelihood of FDA approval for Vioxx, the pharmaceutical giant Merck used flawed methodologies biased toward predetermined results to exaggerate the drug's positive effects.*

You mean exactly like every other Big Pharma company does with every other drug?

> *Merck's manipulation also included a pattern of ghostwriting scientific articles. Internal documents reveal that in 16 of 20 papers reporting on clinical trials of Vioxx, a Merck employee was initially listed as the lead author of the first draft.*

You mean like every other Big Pharma company does with every other drug?

An FDA insider testifying on the agency's failure to quickly recall Vioxx said it was the equivalent of allowing "two to four jumbo jetliners" to crash every week for five years.

The jumbo jetliner analogy is relevant again today, and action by the FDA seems nowhere in sight.

What Ed Dowd has done in this book encourages us to bring our own curiosity and skepticism to the present unprecedented moment, in which new and little-tested Pharma products are being injected into the majority of people on Earth, billions of doses thus far, another 4-million each day, evermore including children, even infants, authorized by an FDA that tried hard to keep Pfizer's clinical trials data secret from the public. To be more accurate, the FDA was perfectly willing and happy to release the information to the public, only they asked for 55-years to do so. To be even more accurate, the FDA (joined by its partner, Pfizer) later petitioned the Court to allow them 75-years to disclose all the information.

Does the fact that the FDA fought so hard to keep the Pfizer safety results secret from the public make you curious?

Well, don't expect any answers from the FDA, now run by a man named Robert Califf. Due to his Big Pharma conflicts, Califf would be ineligible to even serve on an FDA advisory committee, but he is nonetheless the Commissioner, on his third tour of duty at the FDA because, well, the revolving door spins fast, and so does he.

Califf presided over FDA approval of Oxycodone use in children as young as 11, and high-dose hydrocodone drugs, and also the drug Addyl that was later recalled, among several others. For each of these approvals, the FDA overrode its own advisory committee of experts, and Califf gave his approval.

Califf founded a group to do clinical trials, and more than half of his $230 million funding came from... Big Pharma. Almost done. Califf was also paid as a consultant by the 15 biggest Pharma companies. And finally, sure, he was a cheerleader for Vioxx.

So of course he's Commissioner of the FDA.

And don't expect much curiosity or skepticism from CDC Director Rochelle Walensky either. Here she is in March of 2022, displaying her willful gullibility about Pharma claims:

> "I can tell you where I was when the CNN feed came that it was 95% effective, the vaccine. So many of us wanted to be hopeful, so many of us wanted to say, okay, this is our ticket out, right, now we're done. So I think we had perhaps **too little caution and too much optimism** for some good things that came our way. I really do. I think all of us wanted this to be done."

> "Nobody said waning, when you know, oh this vaccine's going to work. Oh well, maybe it'll wear [laughs], it'll wear off."

> "Nobody said what if... [the vaccine is] not as potent against the next variant."

She couldn't predict that vaccines might be less effective against new variants? You mean, like the flu vaccine has done every year for decades?

She couldn't predict that the vaccine would wane, like the flu vaccine has done every year for decades, like even measles vaccines, and tetanus vaccines?

Dr Walensky was once Chief of Infectious Diseases at a hospital where I'd never want to be a patient if she is Chief of Infectious Diseases. But who am I kidding, she's not going back to that hospital; she's more likely to join Pfizer's Board of Directors. Former FDA Commissioner Scott Gottlieb is already there waiting for her.

Dr. Walensky's catchphrase, "Too little caution and too much optimism," could be a lyric in CDC's most famous song, Safe and Effective.

To better understand how the agency always concludes that every vaccine is safe and effective, consider a vaccine so old that all its risk factors are already known: the smallpox vaccine.

From the CDC website, this morning:

> **The smallpox vaccine is safe**, and it is effective at preventing smallpox disease.

Let's see what *safe* means to the CDC:

> *Serious Side Effects of Smallpox Vaccine*
> - *Heart problems*
> - *Swelling of the brain or spinal cord*
> - *Severe skin diseases*
> - *Spreading the virus to other parts of the body or to another person [huh?]*
> - *Severe allergic reaction after vaccination*
> - *Accidental infection of the eye (which may cause swelling of the cornea causing scarring of the cornea, and blindness)*
>
> *The risks for serious smallpox vaccine side effects are greater for:*
> - *People with any three of the following risk factors for heart disease: high blood pressure, high cholesterol, diabetes*
> - *People with heart or blood vessel problems, including angina, previous heart attack or other cardiac problems*
> - *People with skin problems, such as eczema*
> - *Women who are pregnant or breastfeeding*

That's going to be a lot of people risking brain swelling, heart problems and blindness, given that millions of Americans have had heart attacks "or other cardiac problems," 17 million Americans are pregnant or breastfeeding, 31 million have eczema, 34 million have diabetes, 76 million have high cholesterol, and 109 million have high blood pressure.

Describing the people at risk of serious side effects from the safe and effective smallpox vaccine, CDC includes people with a "family history of heart problems."

Do you know hardly anyone who doesn't fit that category?

So, while the CDC definitively states "***The smallpox vaccine is safe***," they then exclude the majority of Americans it might be safe for.

Clearly, the term safe means different things to different experts. Here, for example, is the Mayo Clinic's view of the exact same vaccine the CDC calls safe:

> "It can sometimes cause serious side effects, such as infections in the heart or brain. That's why the vaccine is not given to everyone. Unless there is a smallpox outbreak, ***the risks of the vaccine outweigh the benefits for most people***."

But Gavin, since there's no outbreak of smallpox, why worry about what's on the CDC website? Because... wait for it... the CDC is right now actively promoting that same smallpox vaccine to prevent monkeypox, and nearly a million Americans have already had the pleasure. (I guess it's people with no family history of heart problems.)

And why have just one smallpox vaccine when you can have two?

CDC: 1 dose of vaccine protects against monkeypox, 2nd dose encouraged

—

Experts urge a second dose for full protection, but people who received a single dose of the monkeypox vaccine appeared to be significantly less likely to get sick.

You now know that when CDC says "safe and effective," that's because CDC never met a vaccine they didn't like. It's a belief system – not science.

Carefully reading Ed Dowd's comprehensive analysis of new excess deaths and highest-ever disability among Americans, I was struck that he never once asks us to take his word for anything, and always puts the original source material right in front of us. Even so, and even though he's a brilliant analyst coming to this topic unconflicted, could we really rely upon Ed Dowd's conclusions over the conclusions of the famous and powerful Doctors Fauci, Califf, Bourla and Walensky?

That was my last rhetorical question.

Gavin de Becker

SEEING IS BELIEVING
by Gavin de Becker

I asked Ed Dowd if I could have space in his book for an article about what we saw around the world as mass vaccination commenced. In light of Ed's stunning analysis, it is particularly instructive to look at data for those countries that did not have high numbers of COVID deaths prior to mass vaccination, because they afford the simplest comparison:

1. They had very low rates of death attributed to COVID
2. Then they commenced mass vaccination
3. Then they experienced huge increases in deaths attributed to COVID

South Korea gives us a fast example among many: Prior to their wide rollout of mRNA vaccines, they had almost no COVID deaths. You see that nearly all their COVID deaths occurred <u>after</u> mass vaccination.

Daily New Deaths in South Korea

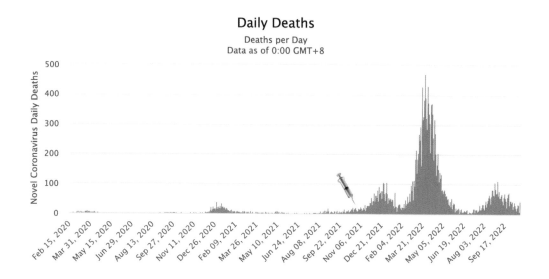

Due to frequent supply problems, South Korea's mass vaccination program really took off after the Third Quarter of 2021 when they borrowed hundreds of thousands of Pfizer doses from Israel. Their COVID deaths soon followed. That wasn't supposed to happen.

In November 2021, President Moon began a massive campaign to push boosters: "The vaccination can be completed only after receiving the third jab." His citizens complied, reaching more than 90% of adults fully vaccinated – the chart shows the COVID deaths that followed.

The same pattern repeats all over the world, and since seeing is believing, I'll pause here and resume in more detail after some quick sample charts...

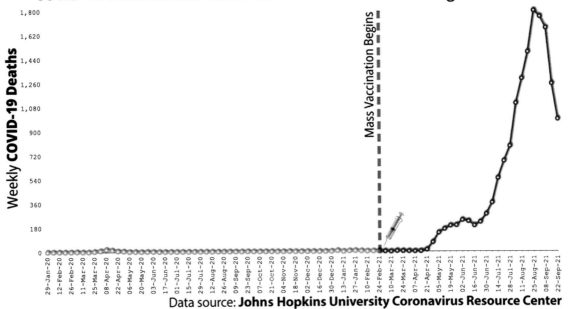

COVID-19 Deaths Before and After Mass Vaccination Program, **Thailand**

Data source: **Johns Hopkins University Coronavirus Resource Center**

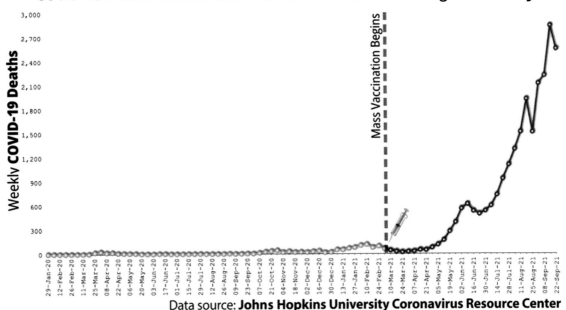

COVID-19 Deaths Before and After Mass Vaccination Program, **Malaysia**

Data source: **Johns Hopkins University Coronavirus Resource Center**

COVID-19 Deaths Before and After Mass Vaccination Program, **Uganda**

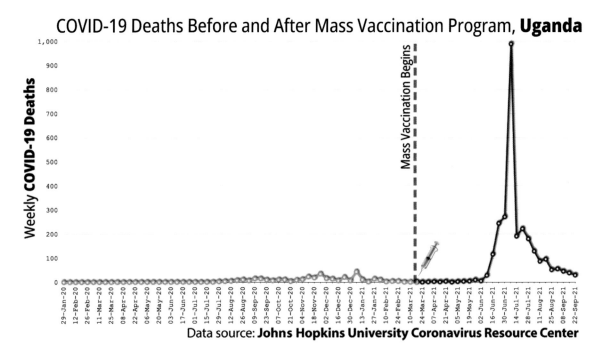

Data source: **Johns Hopkins University Coronavirus Resource Center**

COVID-19 Deaths Before and After Mass Vaccination Program, **Nepal**

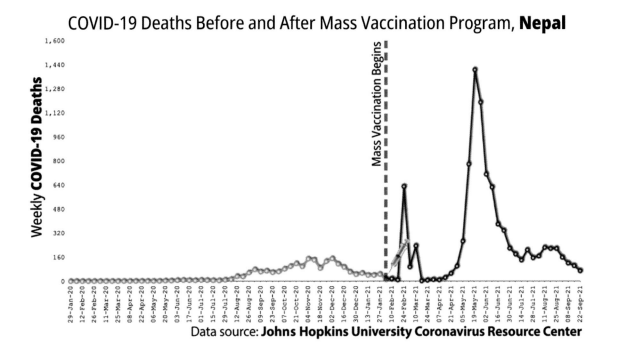

Data source: **Johns Hopkins University Coronavirus Resource Center**

COVID-19 Deaths Before and After Mass Vaccination Program, **Portugal**

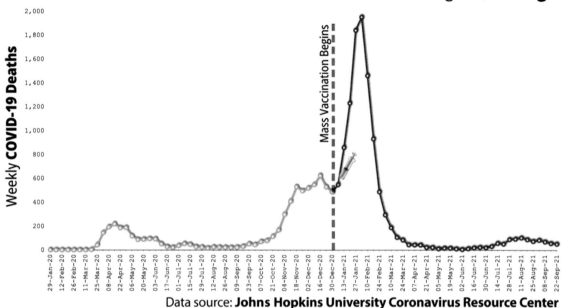

Data source: **Johns Hopkins University Coronavirus Resource Center**

COVID-19 Deaths Before and After Mass Vaccination Program, **Mongolia**

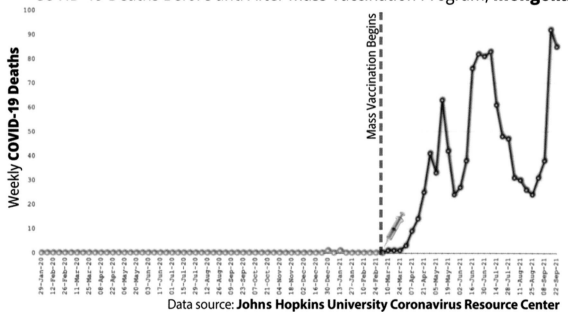

Data source: **Johns Hopkins University Coronavirus Resource Center**

COVID-19 Deaths Before and After Mass Vaccination Program, **Zambia**

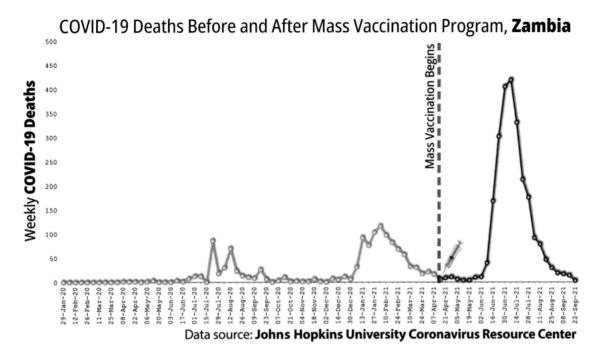

Data source: **Johns Hopkins University Coronavirus Resource Center**

COVID-19 Deaths Before and After Mass Vaccination Program, **Paraguay**

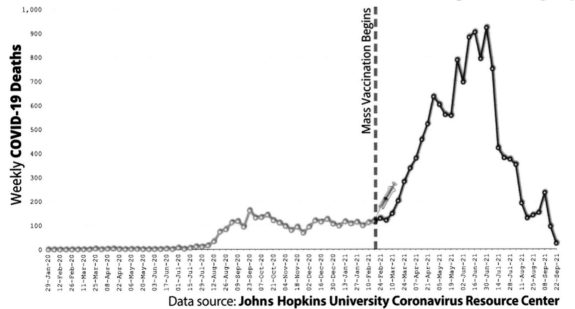

Data source: **Johns Hopkins University Coronavirus Resource Center**

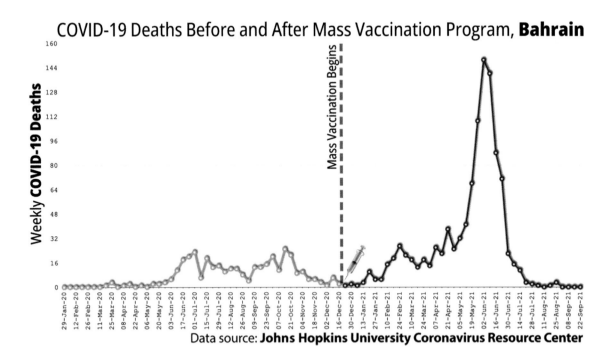

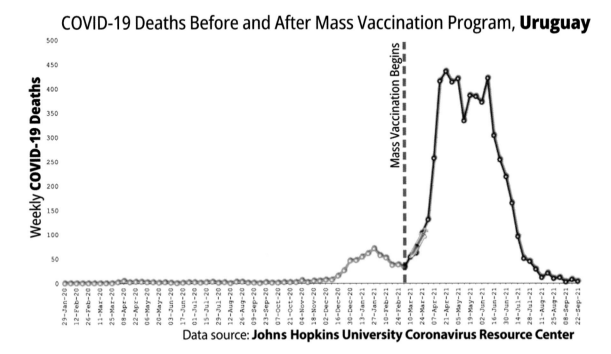

COVID-19 Deaths Before and After Mass Vaccination Program, **Tunisia**

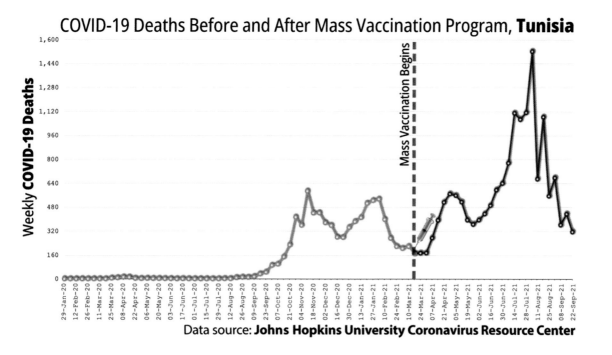

Data source: **Johns Hopkins University Coronavirus Resource Center**

COVID-19 Deaths Before and After Mass Vaccination Program, **Sri Lanka**

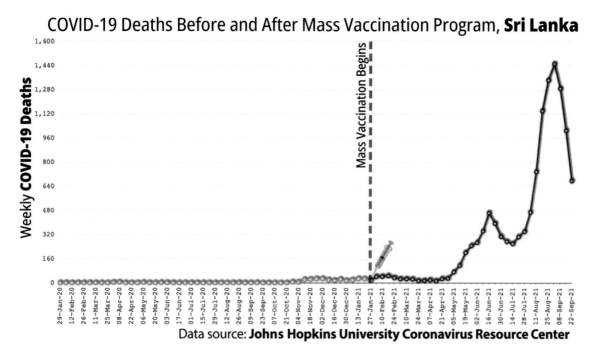

Data source: **Johns Hopkins University Coronavirus Resource Center**

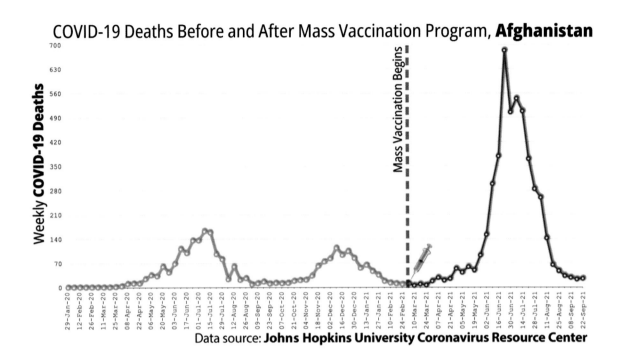

COVID-19 Deaths Before and After Mass Vaccination Program, **Afghanistan**

Data source: **Johns Hopkins University Coronavirus Resource Center**

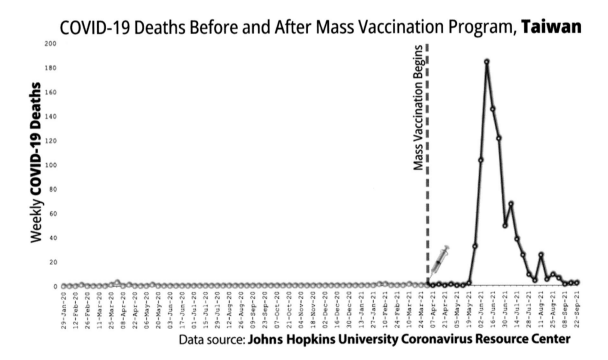

COVID-19 Deaths Before and After Mass Vaccination Program, **Taiwan**

Data source: **Johns Hopkins University Coronavirus Resource Center**

Israel was the world's poster child for Pfizer's vaccine product: Like all these countries, Israel had the majority of its COVID deaths <u>after</u> mass vaccination.

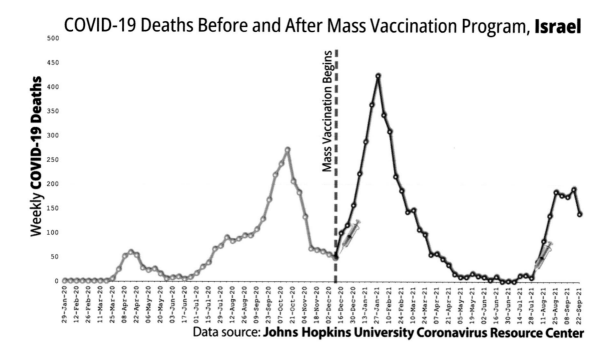

COVID-19 Deaths Before and After Mass Vaccination Program, **Israel**

Data source: **Johns Hopkins University Coronavirus Resource Center**

And finally, Vietnam: They began mass vaccination in March 2021, purchasing five different vaccine products from around the world – and they saw no jump in COVID deaths. However, in early July 2021, the US Government began donating millions of Pfizer and Moderna mRNA vaccines – and that's exactly when Vietnam experienced the massive spike in COVID deaths you see in the chart.

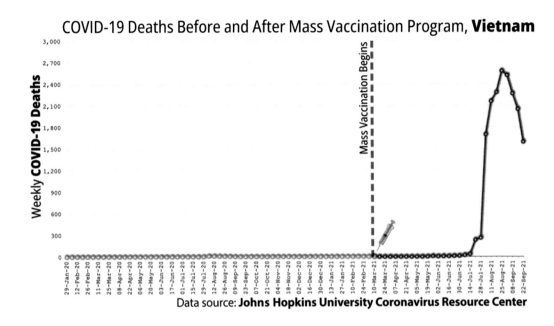

COVID-19 Deaths Before and After Mass Vaccination Program, **Vietnam**

Data source: **Johns Hopkins University Coronavirus Resource Center**

Any way you think about it, those charts should not look like that if vaccination was effective.

Why would so many countries big and small, rich and poor, in different parts of the world, some with congested cities, some sparsely populated, cold weather or hot weather, tropical or desert, high altitude or low altitude, small islands or landlocked – why would they all see increases in COVID deaths after mass vaccination?

That's a question one imagines public health officials and media would be motivated to carefully analyze and answer. Instead, they've been united in keeping such facts out of public discourse. The reality displayed on the graphs you've seen is undeniable, cannot be unseen, and is available to anyone more interested and more industrious than media and government have been.

For curious minds, one explanation to consider is revealed through extensive pre-COVID research establishing that people's immune systems are weakened by some vaccines. Just a few examples among many:

2011 study: Annual vaccination for influenza "may render young children who have not previously been infected with influenza more susceptible to infection with a pandemic influenza virus of a novel subtype."

<u>2013 study</u>: Vaccination may make flu worse if exposed to a second strain [as has been the case with COVID for billions of people].

<u>2018 study</u>: Acute respiratory infections increase following vaccination. This study compared vaccinated people to unvaccinated people.

More recently, a Dutch study of healthcare workers showed a massive increase in COVID infection in the two weeks after the first shot.

Aware of this Danish study, the British Medical Journal published a letter calling for an urgent investigation:

> "Given the evidence of white cell depletion after COVID vaccination and the evidence of increased COVID infection rates shortly after vaccination, the possibility that the two are causally related needs urgent investigation."

The Danish study showed "a 40% increase in infections in the first two weeks after Pfizer-BioNTech vaccination, despite not vaccinating in homes with recent outbreaks," meaning they knew it wasn't because people happened to already be infected at the time they were vaccinated. The 40% number comes up again, in the BMJ letter: "The original Pfizer trial demonstrated a statistically significant 40% increase in suspected COVID."

Looking for a more comfortable answer to the sad riddle, some people might speculate that the deaths you've seen on all those graphs occurred because people became less cautious after vaccination. The British Medical Journal considered and discounted that theory, citing several studies that show increased infections in the weeks after vaccination, and pointing out the example of care home residents, who actually shielded more after vaccination:

"No one is suggesting there was a change of behaviour within care homes. However, *care homes in every corner of the country saw outbreaks from December. What changed?*"

Excellent question. Obvious answer.

If these new Pharma products had been bound by the same laws as all other Pharma products, their TV commercials would have to end with the familiar announcer hurriedly rushing through side effects:

> *COVID vaccines will leave some people more vulnerable to infection and sickness. Some people will experience side effects including cardiac arrest, blood clots, stroke, and sudden death.*

It wouldn't make for a very good sales pitch.

Of course, Pfizer and Moderna didn't need any sales pitch for these vaccines – since the products were developed, ordered, purchased, promoted, defended, indemnified and even mandated by our own government.

COMPENDIUM SAMPLING

This Compendium is a collection of reports on sudden unexpected deaths during 2021 and much of 2022, notably situations in which people collapsed or were found dead in non-hospital settings, meaning they were not under a doctor's care when the medical emergency occurred.

METHODOLOGY

Nearly all incidents in the Compendium were gleaned from news media reports. Though we sometimes encountered conflicting information, the basic issue in these media reports appears accurate: a healthy young person died suddenly and unexpectedly. When possible, we drew upon later reports, since initial or so-called breaking news stories often contain inaccuracy and unreasonable speculation.

When young people collapse and die at sporting events, the collapses are witnessed by spectators and thus more likely to be reported in news stories. Similarly, in small towns, a young athlete might be well known in the community making it more likely that their sudden death will be reported.

In contrast, most children and teenagers dying in their sleep in a big city would not be reported in the news at all. For these and other reasons, the Compendium makes no effort whatsoever to be a comprehensive collection.

In a very few rare instances where age at death was not reported, we estimated based upon appearance in a photo or upon grade level at school.

A SAMPLE, NOT A COMPLETE COLLECTION

While the number of cases described in the Compendium is large, the Compendium is not held out as a complete listing of sudden deaths of healthy young people since 2021 – far from it. Aiming at a comprehensive collection would fail in any event, given that very few of these incidents are reported by national media companies. Rather, the Compendium sought to assemble a sample that is sufficient to offer insight and to support meaningful conclusions.

CRITERIA FOR INCLUSION

There were two fairly obvious criteria to be met before an incident could be considered for inclusion in this Compendium:

1. The incident became known beyond the people and families immediately involved (most often by becoming a news story); or,
2. The incident became known in some other way, such as appearing in the medical or scientific literature.

CRITERIA FOR EXCLUSION

Though there were just two criteria for inclusion of incidents, there were many criteria for *exclusion*. Rumors, cases merely heard about, and second/third/fourth-hand reports were excluded. Also excluded were deaths about which there was any indication that the person was already severely sick at the time of death (for example, in the care of a doctor or hospital). Also excluded: situations in which there is any indication of suicide, homicide, other foul play, drug overdose, accident, etc.

A SAMPLING THAT COMMUNICATES FREQUENCY, HUMANITY, IMPORTANCE, SETTING AND SITUATION

While the Compendium contains many names, ages, and links to news stories, some examples in the book contain images of newspaper headlines and stories. Our purpose is to show, in the least gratuitous and most efficient way possible, that there have been many sudden deaths of healthy young people in 2021/2022, that this is an issue of high public importance, that many sudden deaths have been reported in the news, and that these stories and incidents are widespread.

The headlines themselves often demonstrate that the deaths were shocking and unexpected. Photos often allow the reader to see that the person is young and fit. Photos sometimes communicate that the death would not be expected. Photos depicting athletes in team uniforms, for example, make clear that the young person was, by all appearances, healthy enough to be included in competitive sports, that team management was likely not aware of any health issues for that player, and that death in that setting and situation is unexpected.

The number of headlines or images included in the sampling is a fraction of the total amount available, even a small fraction of the sampling available in this Compendium. In all cases, the content was selected and highlighted from a larger list of cases in support of a specific purpose; for example, it took a certain number of cases to communicate that many healthy people had died in their sleep.

Other images communicated that many famous young people died suddenly in 2021/2022. Still other images communicated that many young medical professionals died suddenly in 2021/2022.

All news headlines and any images included with stories were already public, and newspapers or news outlets had already clearly accomplished their publication goals well in advance of this book, meaning they were not denied any opportunity to publish first. The Compendium team are not aware of any circumstance in which our use could preclude profitable or beneficial use of the same material by news outlets, copyright holders, or anyone else.

Finally, the team found no instances in which images that had been public in news stories were objected to or retracted.

Everyone involved in this book expresses their sadness and condolences about these tragic early deaths.

1 **Helena de Marco**   **5 yrs Died in sleep 8 days post vaccination**   Feb 26, 2022
https://stopcensura.online/bambina-di-5-anni-muore-nel-sonno-8-giorni-dopo-la-prima-dose-pfizer-arresto-cardiaco

2 **Rozalia Spadafora**   **5 yrs Cardiac event / Myocarditis**   Jul 5, 2022
https://www.abc.net.au/news/2022-07-28/act-family-of-five-year-yrs-girl-died-at-canberra-hospital-speak/101276188

3 **Giulia D.L.**   **6 yrs Died unexpectedly in her sleep**   Apr 1, 2022
https://edizionecaserta.net/2022/04/27/bimba-di-6-anni-trovata-morta-sul-divano-di-casa-stava-dormendo/

4 **Unnamed Child**   **8 yrs Died unexpectedly in her sleep**   Sep 6, 2022
https://www.bfmtv.com/police-justice/marmande-un-enfant-de-8-ans-retrouve-mort-a-son-domicile-les-causes-du-deces-totalement-inconnues_AN-202209060305.html

5 **Unnamed Child**   **9 yrs Sudden death at home in front of parents**   Sep 16, 2022
https://www.fanpage.it/attualita/malore-improvviso-in-casa-si-rivela-fatale-bimbo-di-9-anni-muore-davanti-ai-genitori/

6 **Treven Ball**   **10   Sudden death**   Sep 13, 2022
https://www.kgns.tv/2022/09/15/youth-football-player-dies-unexpectedly-after-fulfilling-dream-playing-under-lights/

7 **Kyran Reading**   **10   Sudden death**   Mar 7, 2022
https://www.peterboroughtoday.co.uk/news/people/kyran-reading-memorial-peterborough-community-gathers-as-football-club-pay-tribute-to-boy-10-who-died-suddenly-3609951

8 **Unnamed Child**   **10   Cardiac event**   Apr 19, 2022
https://www.standard.co.uk/news/london/girl-dies-putney-leisure-centre-cardiac-arrest-unexplained-metropolitan-police-b995087.html

9 **Unnamed Child**   **10   Sudden death during gym class**   Aug 2, 2022
https://www.clarin.com/sociedad/conmocion-tartagal-chico-10-anos-murio-clase-educacion-fisica_0_GhfYdDnYnn.html

10 **Unnamed Child**   **11   Cardiac event**   Jul 9, 2022
https://euroweeklynews.com/2022/07/12/11-year-old-cardiac-arrest-dies-pleneuf-val-andre-france/

11 **Ryan Heffernan**   **12   Sudden death at school**   Mar 28, 2022
https://www.echo-news.co.uk/news/21875632.ryan-heffernan-surplices-family-learning-accept-may-no-answers/

12 **Unnamed Child**   **12   Sudden death during run**   Aug 4, 2022
https://www.nzherald.co.nz/nz/baradene-college-student-death-12-year-old-collapses-during-run/XJVCWWML5MWQL2SSIBDLOM75OU/

13 **Chloe Gavazzi**   **12   Sudden death**   Jun 12, 2021
https://www.notizie.it/cloe-giani-giavazzi-morta-a-milano-a-soli-12-anni-addio-alla-giovanissima-promessa-del-tennis/

14 **Braden Fahey**   **12   Sudden death during training**   Aug 7, 2021
https://www.ktvu.com/news/the-death-of-a-middle-school-athlete-is-tragic

15 **Leo Forstenlechner**   **12   Sudden death**   Dec 17, 2021
https://goodsciencing.com/covid/athletes-suffer-cardiac-arrest-die-after-covid-shot/

16 **Carlo Conte**   **12   Cardiac event during run**   Jan 18, 2022
https://www.padovaoggi.it/attualita/carlo-alberto-conte-morto-fiamme-oro-26-gennaio-2022.html

17 **Gemma Caffrey**   **12   Sudden death**   Oct, 2021
https://www.dailyrecord.co.uk/in-your-area/lanarkshire/headteacher-leads-tributes-popular-lanarkshire-25318734

| 18 | **Unnamed Child** | 12 | **Cardiac event 1 day post vaccination** | Feb 20, 2022 |

https://au.news.yahoo.com/investigation-into-12-year-old-boys-death-days-after-covid-vaccine-001210961.html

| 19 | **Mattea Sommerville** | 12 | **Died unexpectedly in her sleep** | Jun, 2022 |

https://euroweeklynews.com/2022/06/10/12-year-old-girl-dies-suddenly/

| 20 | **Althumani Brown** | 13 | **Cardiac event during school field trip** | Jun 1, 2022 |

https://www.cbsnews.com/baltimore/news/baltimore-7th-grader-died-of-natural-causes-on-field-trip-autopsy-finds/

| 21 | **Vanessa Figueiredo** | 13 | **Health declined upon vaccination** | Jan 10, 2022 |

https://dailytelegraph.co.nz/covid-19/watch-the-tragic-story-of-vanessa-martins-figueiredo-brazilian-teen-who-died-after-taking-covid-jab/

| 22 | **Chino Nsofor** | 13 | **Sudden death during training** | Jun 28, 2021 |

https://www.reviewjournal.com/local/coroner-reveals-cause-of-legacy-high-school-football-players-death-2499882/

| 23 | **Samuel Akwasi** | 13 | **Cardiac event during game** | May 7, 2022 |

https://www.mirror.co.uk/sport/football/news/footballer-dies-medical-emergency-nottingham-26905775

| 24 | **Joshua Henry** | 13 | **Brain bleed 90 minutes post vaccination** | Oct 4, 2021 |

https://rupreparing.com/news/2021/11/23/joshua-henry-14-year-old-boy-dies-from-massive-brain-bleed-hours-after-receiving-second-pfizer-covid-19-vaccine

| 25 | **Jacob Clynick** | 13 | **Cardiac event 3 days post vaccination** | Jun 16, 2021 |

https://nypost.com/2021/07/05/michigan-boy-dies-in-his-sleep-three-days-after-getting-vaccine/

| 26 | **Marco Benitez** | 13 | **Sudden death during game** | Oct 22, 2021 |

https://www.kusi.com/memorial-held-for-oceanside-middle-school-student-who-died-suddenly/

| 27 | **Sophia Fenner** | 13 | **Sudden death** | Jan 17, 2022 |

https://navarrepress.com/friends-rally-to-help-family-of-13-year-old-girl-who-died-suddenly/

| 28 | **Jack O'Drain** | 13 | **Cardiac event** | Jan 1, 2022 |

https://www.gofundme.com/f/the-odrain-family

| 29 | **Jorge Mezcua** | 13 | **Sudden death** | Apr 3, 2022 |

https://euroweeklynews.com/2022/04/05/young-footballer-died-suddenly-cadiz/

| 30 | **Unnamed Child** | 13 | **Cardiac event on the playground** | Mar 28, 2022 |

https://euroweeklynews.com/2022/03/28/sudden-death-of-13-year-old-pupil-in-playground-at-the-ies-sierra-de-mijas/

| 31 | **Sheikh Shaheeda** | 13 | **Cardiac event during class** | Sep 7, 2022 |

http://timesofindia.indiatimes.com/articleshow/94063699.cms

| 32 | **Nico** | 14 | **Sudden death** | Jan, 2022 |

https://www.merkur.de/lokales/fuerstenfeldbruck/olching-ort29215/bayern-fasching-olching-vorfall-tot-91226280.html

| 33 | **Karol Setniewski** | 14 | **Sudden death** | Dec 17, 2021 |

https://www.tvp.info/57514609/nie-zyje-mlody-pilkarz-znicza-pruszkow-karol-seta-setniewski-mial-13-lat-byl-kandydatem-do-akademii-legii-warszawa

| 34 | **Wouter Betjes** | 14 | **Sudden death** | Dec 4, 2021 |

https://www.ad.nl/binnenland/wouter-14-zakt-in-elkaar-op-hockeyveld-en-overlijdt-school-en-club-in-rouw~a064d150/

35	**Talina Rampersad**	**14**	**Sudden death**	Jul 17, 2022
	https://www.cbc.ca/news/canada/manitoba/talina-rampersad-husack-death-autopsy-wait-1.6534329			
36	**Giada Furlanut**	**14**	**Blood clot**	Dec 5, 2021
	https://pledgetimes.com/giada-dies-at-the-age-of-14-after-having-been-ill-at-school/			
37	**Aoibhe Byrne**	**14**	**Sudden death**	Apr 27, 2022
	https://www.irishmirror.ie/news/irish-news/monaghan-community-state-shock-following-26824075			
38	**Unnamed Child**	**14**	**Cardiac event during training**	Sep 7, 2022
	https://www.ansa.it/english/news/general_news/2022/09/07/boy-14-dies-5-days-after-cardiac-arrest-on-soccer-pitch_f602d1d8-1d35-4c6a-976d-1e30fc8c0548.html			
39	**Milagros Santiso**	**14**	**Cardiac event**	Jul 30, 2022
	https://aleteia.org/2022/08/20/14-year-old-girl-dies-on-a-retreat-her-mothers-words-of-faith-will-inspire-you/			
40	**Unnamed Child**	**14**	**Sudden death during game**	Aug 4, 2021
	https://www.insidehalton.com/news-story/10450254-milton-boy-14-dies-after-collapsing-in-parking-lot-while-playing-basketball/			
41	**Jack Pollock**	**14**	**Sudden death**	Sep 29, 2022
	https://www.edinburghlive.co.uk/news/edinburgh-news/tributes-loving-edinburgh-schoolboy-who-25159335			
42	**Unnamed Child**	**15**	**Cardiac event during training**	Apr 29, 2022
	https://www.francebleu.fr/infos/faits-divers-justice/un-adolescent-de-15-meurt-d-un-malaise-cardiaque-pendant-un-entrainement-de-football-a-contrexeville-1651307095			
43	**Unnamed Child**	**15**	**Sudden death**	Aug 3, 2022
	https://www.mmegi.bw/sports/mexican-girls-starlet-collapses-dies-in-training/news			
44	**DeVonte Mumphrey**	**15**	**Sudden death during game**	Feb 8, 2022
	https://www.newsweek.com/devonte-mumphrey-texas-high-school-basketball-game-alto-texas-1677533			
45	**Brian Saes**	**15**	**Cardiac event while cycling**	Jul 21, 2021
	https://g1.globo.com/sc/santa-catarina/noticia/2021/07/22/adolescente-morre-durante-passeio-de-bicicleta-em-sc-era-um-menino-muito-feliz-diz-tia.ghtml			
46	**Isaiah Banks**	**15**	**Sudden death**	Jul 11, 2021
	https://www.fox5atlanta.com/news/norcross-high-school-mourns-football-player-isaiah-banks-death			
47	**Stephen Sylvester**	**15**	**Sudden death**	Aug 2, 2021
	https://www.theoaklandpress.com/2021/08/10/catholic-central-mourning-the-loss-of-football-track-athlete-stephen-sylvester/			
48	**Pedro Oliveira**	**15**	**Sudden death**	Jan 11, 2022
	https://www.thegatewaypundit.com/2022/01/15-year-old-soccer-player-brazil-dies-cardiac-arrest-following-national-football-cup-tournament-game/			
49	**Preston Settles**	**15**	**Cardiac event during game**	Feb 5, 2022
	https://www.cbsnews.com/boston/news/preston-settles-funeral/			
50	**Elias Georgakopoulos**	**15**	**Sudden death 3 days post vaccination**	Oct, 2021
	https://www.eventiavversinews.it/elias-georgakopoulos-15-anni-perfettamente-sano-muore-3-giorni-dopo-aver-ricevuto-il-vaccino-pfizer-covid-19-il-fratello-racconta/			
51	**Danylo Nobre**	**15**	**Became sick 18 days post vaccination**	Mar 3, 2022
	https://community.covidvaccineinjuries.com/danylo-15-year-old-died-after-pfizer-induced-brainstem-encephalitis/			

52   **Asheley Garcia**   **15**   **Stroke 5 days post vaccination**   Mar 13, 2022
https://diariodevallarta.com/en/la-muerte-de-asheley-carbajal-garcia-si-no-se-hubiera-
vacunado-estaria-viva/

53   **Matteo Pietrosanti**   **15**   **Sudden death during training**   Mar 3, 2022
https://pledgetimes.com/latina-matteo-pietrosanti-dies-at-the-age-of-15-in-front-of-his-
mothers-eyes/

54   **Unnamed Child**   **15**   **Sudden death 2 days post vaccination**   Jun 7, 2021
https://www.pressdemocrat.com/article/news/county-officials-social-media-posters-
spar-over-boys-death/

55   **Nico Holguin**   **15**   **Sudden death**   Mar 21, 2022
https://www.infobierzo.com/fallece-en-leon-el-joven-nico-holguin-de-15-anos-que-fue-
reanimado-en-el-parque-del-plantio-de-ponferrada/669304/

56   **Carmyne Payton**   **15**   **Sudden death during game**   Nov 18, 2021
https://people.com/sports/15-year-old-boy-collapses-dies-basketball-tryouts-long-
island-new-york/

57   **Unnamed Child**   **16**   **Cardiac event during game**   Feb 25, 2022
https://primeraplana.mx/archivos/858300

58   **Unnamed Child**   **16**   **Sudden death during training**   Sep 6, 2022
https://www.cbsnews.com/baltimore/news/randallstown-high-school-student-dies-after-
medical-emergency-at-football-practice/?intcid=CNM-00-10abd1h

59   **Anna Burns**   **16**   **Cardiac event at cross-country meet**   Sep 13, 2022
https://www.amherstbulletin.com/Anna-Burns-remembered-by-coaches-
teammates-48204979

60   **Henry Farmer**   **16**   **Sudden death**   Apr 12, 2022
https://patch.com/connecticut/darien/darien-high-hockey-player-remembered-great-
teammate-hard-worker

61   **Ernesto Ramirez, Jr.**   **16**   **Cardiac event 5 days post vaccination** Apr 24, 2021
https://circleofmamas.com/health-news/grieving-father-ernest-ramirez-shares-
heartbreaking-story-of-his-teen-sons-death-5-days-after-pfizer-vaccine/

62   **Joshua Johnson**   **16**   **Sudden death**   Jun, 2021
https://dailyprogress.com/community/greenenews/news/memorial-to-remember-
joshua-johnson-on-thursday-june-10/article_98a9194c-c942-11eb-b70d-d3d85f6333a1.
html

63   **Ethan Trejo**   **16**   **Sudden death**   Jun 25, 2021
https://www.cincinnati.com/story/news/2021/06/25/teen-dies-after-medical-incident-
princeton-high-school-field/5344293001/

64   **Devon DuHart**   **16**   **Sudden death**   Jul 24, 2021
https://www.thv11.com/article/sports/little-rock-central-death-devon-duhart/91-
82359c68-9a8e-4611-90fb-cf679ce1ee72

65   **Jascha Zey**   **16**   **Sudden death**   Jul 28, 2021
https://www.sportfreunde-eisbachtal.de/ploetzlich-und-viel-zu-frueh-die-eisbaeren-
familie-trauert-um-u19-spieler-jascha-zey

66   **Nathan Esparza**   **16**   **Cardiac event**   Jul 13, 2021
https://scvnews.com/castaic-high-school-brings-grief-counselors-on-campus-after-
student-death/

67   **Jamarcus Hall**   **16**   **Sudden death**   Aug 11, 2021
https://www.wlbt.com/2021/11/09/16-year-old-mississippi-football-player-dies/

68   **Antonio Hicks**   **16**   **Sudden death**   Sep 28, 2021
https://www.nfldraftdiamonds.com/2021/10/antonio-hicks/

69	**Jony Lopez**	16	**Cardiac event**	Nov 11, 2021
	https://radioconcierto.com.py/2021/11/12/futbolista-infarto-durante-practica/			
70	**Valentin Rodionov**	16	**Sudden death during game**	Nov 28, 2021
	https://www.rt.com/sport/541525-russian-ice-hockey-young-star-dies/?s=09			
71	**Mosheur Rahman**	16	**Adverse events upon vaccination**	Aug 30, 2021
	https://fleekus.com/b/antonio-silva-fkzle/article/mosheur-rahman-healthy-16-year-old-boy-dies-shortly-after-receiving-the-moderna-covid-19-vaccine-family-seeks-justice-fgtyq			
72	**Melanie Macip**	16	**Cardiac event 5 days post vaccination**	Aug 7, 2021
	https://twitter.com/veritebeaute/status/1439363821370494977			
73	**Sofia Benharira**	16	**Cardiac event / Blood clot**	Sep 21, 2021
	https://www.australiannationalreview.com/covid-19-deaths-and-injuries/sofia-benharira-16-years-old-died-from-pfizer-vaxxine/			
74	**Isabelli Valentim**	16	**Sudden death**	Sep 2, 2021
	https://thecovidblog.com/2021/09/23/isabelli-borges-valentim-16-year-old-brazilian-girl-develops-blood-clots-dead-eight-days-after-first-pfizer-mrna-injection/			
75	**Kamrynn Thomas**	16	**Cardiac event soon post vaccination**	Mar 30, 2021
	https://healthimpactnews.com/2021/16-year-old-wisconsin-girl-dead-following-2-doses-of-the-experimental-pfizer-covid-injections/			
76	**Amy Forde**	16	**Died unexpectedly in her sleep**	Nov 23, 2021
	https://www.mylondon.news/news/east-london-news/healthy-girl-16-died-out-24349492			
77	**Lee Nolan**	16	**Sudden death**	May 13, 2022
	https://extra.ie/2022/05/19/news/irish-news/community-in-wexford-deeply-saddened-following-death-of-incredibly-kind-young-student			
78	**Kevin Amaya**	16	**Cardiac event during game**	Jul 10, 2022
	https://www.periodicoequilibrium.com/adolescente-de-16-anos-muere-de-un-infarto-durante-partido-de-futbol/			
79	**Unnamed Child**	16	**Sudden death during training**	Jun 8, 2022
	https://www.wtkr.com/news/bayside-high-school-student-athlete-dies-after-collapsing-during-conditioning			
80	**Owen Cotty**	16	**Cardiac event while playing Frisbee**	Aug 6, 2022
	https://patch.com/pennsylvania/lowerprovidence/montco-teen-dies-cardiac-arrest-while-playing-frisbee			
81	**Gianluca Schettino**	16	**Sudden death**	May 15, 2022
	https://napoli.occhionotizie.it/gragnano-ragazzo-morto-gianluca-funerali-16-anni/			
82	**Quentin Watson**	16	**Sudden death**	Aug 18, 2022
	https://dailyvoice.com/pennsylvania/montgomery/news/beloved-norristown-area-high-school-student-dies-suddenly-at-16/841307/			
83	**D.J.**	16	**Sudden death during game**	Sep 19, 2021
	https://www.telegraf.rs/english/3393101-young-football-players-pulse-returned-briefly-but-he-couldnt-be-saved-and-passed-away-in-seconds			
84	**Cameran Wheatley**	17	**Sudden death during game**	Feb 8, 2022
	https://abc7chicago.com/cameran-wheatley-bremen-high-school-basketball-christ-hospital/11549978/			
85	**Bailey Munro**	17	**Sudden death**	Jul 21, 2022
	https://www.pressandjournal.co.uk/fp/news/inverness/4576309/death-of-17-year-old-in-inverness-bailey-matheson-munro/			

| 86 | **Ali Muhammad** | 17 | **Died unexpectedly in his sleep** | Sep 8, 2022 |

https://www.westernjournal.com/seemingly-healthy-17-year-old-football-player-dies-sleep-will-forever-thoughts/

| 87 | **Shruti Soni** | 17 | **Cardiac event** | Sep 30, 2021 |

https://www.freepressjournal.in/bhopal/madhya-pradesh-teen-players-death-due-to-cardiac-arrest-triggers-concern

| 88 | **Miguel Lugo** | 17 | **Sudden death during training** | Mar 3, 2021 |

https://www.recordonline.com/story/sports/high-school/2021/03/02/wallkill-football-player-miguel-lugo-dies-after-practice-on-monday/6894023002/

| 89 | **Andrew Roseman** | 17 | **Sudden death** | Jul 15, 2021 |

https://www.phillyvoice.com/red-land-baseball-player-dies-andrew-roseman-pennsylvania-york-county/

| 90 | **Donadrian Robinson** | 17 | **Sudden death** | Aug 29, 2021 |

https://www.wistv.com/2021/09/04/he-would-love-it-donadrian-robinsons-family-reacts-tribute-wj-keenan-high-school/

| 91 | **Sean Hartman** | 17 | **Myocarditis 4 days post vaccination** | Sep 27, 2021 |

https://www.lifesitenews.com/news/all-he-wanted-to-do-was-play-hockey-grieving-dad-says-pfizer-shot-killed-his-17-year-old-son/

| 92 | **Dylan Rich** | 17 | **Cardiac event during game** | Sep 7, 2021 |

https://www.bbc.com/news/uk-england-nottinghamshire-58462925

| 93 | **Krystian Kozek** | 17 | **Sudden death** | Oct 31, 2021 |

https://gol24.pl/nie-zyje-krystian-kozek-17letni-zawodnik-wisloka-strzyzow/ar/c2-15879099

| 94 | **Nathan Rogalski** | 17 | **Sudden death** | Jan 23, 2022 |

https://www.oklahoman.com/story/sports/high-school/baseball/2022/01/23/deer-creek-high-school-baseball-player-dies-sudden-illness-nathan-rogalski/6632572001/

| 95 | **Viggo Sorensen** | 17 | **Cardiac event** | Jan 27, 2022 |

https://www.facebook.com/GEMSWellingtonAcademy.AlKhail.Dubai/photos/a.215399978658681/1784485951750068/?type=3

| 96 | **Philip Laster Jr.** | 17 | **Sudden death during training** | Aug 1, 2022 |

https://www.wapt.com/article/brandon-high-school-football-player-dies-district-confirms/40777412#

| 97 | **Cesar Vasquez** | 17 | **Sudden death** | Aug 2, 2022 |

https://www.azcentral.com/story/sports/high-school/2022/08/04/peoria-centennial-football-community-rocked-death-player/10234394002/

| 98 | **Shubham Chopde** | 17 | **Cardiac event** | Jul 29, 2022 |

https://indianexpress.com/article/cities/pune/pune-student-dies-of-suspected-cardiac-arrest-on-college-trip-to-raireshwar-fort-8060438/

| 99 | **Gwen Casten** | 17 | **Died unexpectedly in her sleep** | Jun 13, 2022 |

https://www.cbsnews.com/chicago/news/gwen-casten-congressman-sean-casten-daughter-death-sudden-cardiac-arrhythmia-heart-condition/

| 100 | **Adam Ali** | 17 | **Sudden death** | Sep, 2021 |

https://www.birminghammail.co.uk/news/midlands-news/a-truly-gentle-soul-tributes-21763398

| 101 | **Unnamed Child** | 17 | **Sudden death** | Sep, 2022 |

https://www.leggo.it/italia/milano/morto_casa_17_anni_varazze_cosa_successo-6908139.html

| 102 | **Tyler Erickson** | 17 | **Sudden death during training** | Sep 12, 2022 |

https://www.wjhg.com/2022/09/13/community-mourns-death-holmes-county-athlete/

103	**Ali Muhamad**	**17**	**Died unexpectedly in his sleep**	Sep 8, 2022

https://unionnewsdaily.com/sports/rahway-varsity-football-player-dies-in-his-sleep-team-to-honor-him-this-season

104	**Kooper McCabe**	**17**	**Sudden death**	Aug 30, 2022

https://www.mansfieldnewsjournal.com/story/news/2022/09/02/galion-student-athlete-remembered-after-his-unexpected-death/65466123007/

105	**Rohan Cosgriff**	**17**	**Sudden death**	Jul 29, 2022

https://www.racing.com/news/2022-07-31/news-industry-ballarat-mourns-cosgriff

106	**Alessia De Nadai**	**17**	**Sudden death**	Jul 4, 2022

https://www.ilparagone.it/cronaca/alessia-muore-a-17-anni-il-malore-i-due-interventi-alla-testa-e-la-fine-origine-sconosciuta/

107	**Paola Alcantara**	**17**	**Cardiac event during lunch break**	May 26, 2022

https://euroweeklynews.com/2022/05/27/17-year-old-girl-cardiac-arrest/

108	**Eitan Force**	**17**	**Sudden death during game**	Sep 21, 2022

https://www.msn.com/en-us/news/us/high-school-student-dies-during-flag-football-game/ar-AA127NPi?li=BBnb7Kz

109	**Blake Barklage**	**17**	**Cardiac event during game**	Oct 30, 2021

https://6abc.com/blake-barklage-death-lasalle-college-high-school-montgomery-county/11187002/

110	**Estefania Arroyo**	**18**	**Sudden death**	Sep 25, 2021

https://www.cuestonian.com/cuesta-college-student-athlete-cause-of-death-remains-a-mystery/

111	**Adrien Sandjo**	**18**	**Cardiac event during game**	Dec 22, 2021

https://sportnewsafrica.com/en/at-a-glance/italy-adrien-sandjo-cameroonian-footballer-dies-after-a-cardiac-arrest/

112	**Alberto Torrecilla**	**18**	**Sudden death**	Jan 20, 2022

https://www.eventiavversinews.it/madrid-muore-per-arresto-cardiaco-improvviso-giovedi-20-gennaio-il-calciatore-alberto-torrecilla-delle-giovanili-del-club-deportivo-avance/

113	**Kasey Turner**	**18**	**Blood clot 14 days post vaccination**	Feb 27, 2021

https://www.yorkshirepost.co.uk/health/family-of-teenager-who-died-after-having-astrazeneca-jab-pay-tribute-to-cheeky-daughter-3625541

114	**Jacob Downey**	**18**	**Cardiac event**	Sep 29, 2021

https://www.thepeterboroughexaminer.com/sports/hockey/2021/10/01/a-brilliant-kid-on-and-off-the-ice-and-in-every-sport-he-played.html

115	**Emmanuel Antwi**	**18**	**Sudden death during game**	Mar 22, 2021

https://www.cbsnews.com/sacramento/news/kennedy-high-emmanuel-antwi-collapses-during-game-dies/

116	**Nikita Sidorov**	**18**	**Sudden death during game**	Apr 4, 2021

https://www.rt.com/sport/520142-russian-football-player-death-znamya-truda/

117	**Victor Hegedus**	**18**	**Sudden death during training**	Jun 21, 2021

https://www.budapestherald.hu/sport/2021/06/26/an-18-year-old-hungarian-football-player-collapsed-and-died-during-training/

118	**Jack Alkhatib**	**18**	**Cardiac event during training**	Aug 24, 2021

https://www.wistv.com/2021/08/28/dutch-fork-high-school-honors-life-jack-alkhatib-with-memorial/

119	**Amir Aiana**	**18**	**Cardiac event during game**	Jan 11, 2022

https://www.milanotoday.it/cronaca/morto-oratorio-san-giustino.html

120 **Mateo Hernandez** 18 **Sudden death** Jan 11, 2022
https://www.eventiavversinews.it/muore-martedi-11-gennaio-a-18-anni-per-malore-improvviso-il-portiere-spagnolo-mateo-hernandez/

121 **Lucas Dias** 18 **Sudden death during workout** Jan 15, 2022
https://www.tribunadecianorte.com.br/ultimas-noticias/cianorte-policia-aguarda-laudo-para-definir-causa-da-morte-de-jovem-em-academia/

122 **Leo McBride** 18 **Sudden death** Oct 24, 2021
https://www.irishmirror.ie/news/irish-news/community-pays-tribute-teenager-heart-25294844

123 **Alessia Raiciu** 18 **Died unexpectedly in her sleep** Aug 15, 2022
https://tinyurl.com/5xnvj5ex

124 **Cameron Milton** 18 **Died unexpectedly in his sleep** Mar 29, 2022
https://kqeducationgroup.com/tributes-paid-to-bolton-school-student-cameron-milton/

125 **Valentina Yazenok** 18 **Neurological events days post vaccination** Jan 11, 2022
https://kam24.ru/news/main/20220112/86541.html

126 **Dani Gómez** 18 **Sudden death** Aug 23, 2022
https://aragondigital.es/deportes/2022/08/23/fallece-dani-gomez-jugador-de-18-anos-del-penas-huesca-de-baloncesto/

127 **Filippo Venezia** 18 **Sudden death** Aug 19, 2022
https://www.ilmessaggero.it/sport/altrisport/filippo_dalla_venezia_morto_casa_rugby_mogliano-6878936.html

128 **Avery Gilbert** 18 **Sudden death** Aug 10, 2022
https://patch.com/illinois/grayslake/s/id0uy/football-player-dies-after-collapsing-on-college-campus-in-deerfield

129 **Jessica Matthews** 18 **Sudden death during game** Jun 15, 2022
https://www.dailymaverick.co.za/article/2022-06-17-western-province-and-maties-hockey-player-collapses-dies-on-field/

130 **Alejandro Candela** 18 **Sudden death** Jun 14, 2022
https://www.lavozdelanzarote.com/actualidad/sociedad/fallece-repentinamente-el-joven-alejandro-candela-una-de-las-promesas-de-la-natacion-en-lanzarote_128553_102_amp.html

131 **Tobiloba Taiwo** 18 **Sudden death during game** Feb 21, 2022
https://mndaily.com/271256/news/breaking-umn-student-dies-unexpectedly-at-recwell-center/

132 **Camilla Canepa** 18 **Blood clot days post vaccination** Jun 10, 2021
https://genova.repubblica.it/cronaca/2021/06/10/news/e_morta_camilla_canepa_la_18enne_ligure_vaccinata_con_astrazeneca-305366834/

133 **Davide Bristot** 18 **Died in sleep weeks post vaccination** Jul 14, 2021
https://www.true-news.it/facts/davide-bristot-morto-a-18-anni-perche-si-indaga-sul-vaccino

134 **Kacper Zabrzycki** 18 **Sudden death** Aug 22, 2021
https://www.o2.pl/sport/nie-zyje-kacper-zabrzycki-mial-tylko-18-lat-jestesmy-wstrzasnieci-6675225822878368a

135 **Aidan Kaminska** 19 **Sudden death** May 30, 2022
https://www.the-sun.com/news/5475432/aidan-kaminska-umass-lacrosse-player-died/

136 **Unnamed Person** 19 **Cardiac event during run** Sep 6, 2022
https://www.linfokwezi.fr/un-gamin-de-19-ans-emporte-par-une-crise-cardiaque-pendant-un-footing/

137 **Keanu Breurs**   **19**   **Sudden death during training**   Jan 12, 2021
https://www.hln.be/beveren/dinsdag-nog-op-training-woensdag-plots-overleden-voetbalclub-svelta-in-rouw-na-verlies-van-talentvolle-en-betrokken-speler-keanu-19~a7e8b759/

138 **Kamila Label-Farrell**   **19**   **Sudden death during run**   Jun 9, 2021
https://www.baytoday.ca/obituaries/lebel-farrell-kamila-3884874

139 **Aidan Price**   **19**   **Sudden death**   Jun 20, 2021
https://www.carleton.edu/farewells/aidan-price-24/

140 **Whitnee Abriska**   **19**   **Cardiac event**   Jul 26, 2021
https://www.dhnet.be/sports/sport-regional/liege/2021/07/26/la-joueuse-du-femina-vise-decede-subitement-a-lage-de-19-ans-PV6FI7R2QNBZTMUTUNNINTPXBY/

141 **Marco Tampwo**   **19**   **Cardiac event**   Aug 15, 2021
https://www.thesun.co.uk/sport/football/15892036/footballer-marco-tampwo-dead-heart-attack-covid/

142 **Tirrell Williams**   **19**   **Stroke during training**   Aug 4, 2021
https://www.fourstateshomepage.com/news/local-news/fscc-football-player-passes-away/

143 **Quandarius Wilburn**   **19**   **Sudden death during training**   Aug 8, 2021
https://www.nbcnews.com/news/us-news/virginia-union-university-football-player-dies-after-collapsing-during-practice-n1276410

144 **Sebastiaan Bos**   **19**   **Sudden death**   Sep 11, 2021
https://goodsciencing.com/covid/athletes-suffer-cardiac-arrest-die-after-covid-shot/

145 **Anna Biktimirova**   **19**   **Sudden death**   Dec 1, 2021
https://www.rt.com/sport/541846-arina-biktimirova-taekwondo-perm-death/

146 **Zachary Icenogle**   **19**   **Died unexpectedly in his sleep**   Dec 26, 2021
https://patch.com/illinois/plainfield/long-live-ice-teens-unexpected-death-has-community-mourning

147 **Volodymyr Salo**   **19**   **Cardiac event hours post vaccination**   Sep 13, 2021
https://expose-news.com/2021/09/29/19-year-old-ukrainian-student-gets-pfizer-vaccine-behind-his-familys-back-dies-seven-hours-later/

148 **Inês Rafael**   **19**   **Sudden death 5 days post vaccination**   Aug, 2021
https://www.cmjornal.pt/portugal/detalhe/universitaria-morre-cinco-dias-apos-a-vacina-da-covid-19-em-vieira-do-minho?ref=HP_PrimeirosDestaques

149 **Simone Scott**   **19**   **Cardiac event / Myocarditis**   Jun 11, 2021
https://www.fox19.com/2021/06/16/mason-high-school-graduate-remembered-kind-talented-after-mysterious-illness-takes-her-life/

150 **Jaysley-Louise Beck**   **19**   **Sudden death**   Dec 15, 2021
https://www.wiltshirelive.co.uk/news/wiltshire-news/familys-heart-breaking-tribute-beautiful-6534286

151 **Matthias Pedersen**   **19**   **Sudden death**   Apr 24, 2022
https://sport.tv2.dk/haandbold/2022-04-24-u20-landsholdsspiller-matthias-birkkjaer-er-pludselig-doed

152 **Teun Elbers**   **19**   **Sudden death while hiking**   Aug 5, 2022
https://www.telegraaf.nl/nieuws/908171433/voetbalclub-oss-rouwt-om-overleden-teun-elbers-19

153 **Stephanie Ming**   **19**   **Stroke**   Jun 19, 2022
https://www.theborneopost.com/2022/06/23/sarawak-sukma-shooter-stephanie-dies-at-19/

154 **Giuseppe Gallina**   **19**   **Sudden death during game**   Feb 12, 2022
https://41esimoparallelo.it/2022/02/22/tragedia-a-statte-e-giuseppe-gallina-il-19enne-morto-durante-la-partita-di-calcetto-sotto-gli-occhi-dei-compagni-aperta-uninchiesta/37/

155 **Rittika Das**   **19**   **Cardiac event during workout**   Aug 9, 2022
https://timesofindia.indiatimes.com/city/kolkata/19-year-old-falls-ill-at-gym-dies-in-kolkata/articleshow/93465703.cms

156 **Brazil Walsh**   **20**   **Sudden death**   Apr 24, 2022
https://www.leeds-live.co.uk/news/leeds-news/woman-20-died-rare-brain-24213876

157 **Derek Gray**   **20**   **Cardiac event during game**   Jul 24, 2022
https://www.msn.com/en-us/sports/ncaabk/college-basketball-star-derek-gray-dead-at-20/ar-AA1072PN

158 **Lily Ernst**   **20**   **Sudden death**   Jul 27, 2022
https://unipanthers.com/news/2022/7/28/womens-swimming-and-diving-uni-mourns-the-loss-of-swimming-student-athlete-lily-ernst.aspx

159 **Eli Palfreyman**   **20**   **Sudden death during half-time**   Aug 30, 2022
https://kitchener.ctvnews.ca/ayr-centennials-missing-the-heart-and-soul-of-our-team-after-captain-s-death-1.6052332

160 **Archie Bruce**   **20**   **Sudden death**   Aug 14, 2021
https://www.mirror.co.uk/sport/rugby-league/breaking-archie-bruce-dead-20-18956732

161 **Christian Blandini**   **20**   **Cardiac event**   Sep 9, 2021
https://freewestmedia.com/2021/09/16/sudden-death-of-young-italian-athlete-and-the-conspiracy-of-silence/

162 **Kim Kyeong-Bo**   **20**   **Sudden death**   Nov 8, 2021
https://www.gamespot.com/articles/professional-overwatch-player-kim-alarm-kyeong-bo-dies-at-20/1100-6497812/

163 **Vinicius Freitas**   **20**   **Cardiac event**   Dec 3, 2021
https://www.meiahora.com.br/esportes/2021/12/6290244-jovem-com-passagem-por-clube-carioca-morre-de-infarto-aos-20-anos.html

164 **Ali Arabzada**   **20**   **Sudden death**   Dec 17, 2021
https://tolonews.com/sport-175926

165 **Tatjana Jagodic**   **20**   **Blood clot / Brain bleed**   Sep, 2021
https://slovenia.postsen.com/local/28539/It-will-be-a-year-since-the-death-of-Katja-for-whom-the-vaccine-against-covid-19-was-fatal.html

166 **Regan Lewis**   **20**   **Cardiac event 1 day post vaccination**   Sep 27, 2022
https://citizenfreepress.com/breaking/healthy-young-student-is-dead-one-day-after-covid-vaccine/

167 **Djouby Laura**   **20**   **Cardiac event**   Aug 11, 2022
https://monewsguyane.com/2022/08/12/un-footballeur-de-lusc-roura-meurt-dun-arret-cardiaque/

168 **Andrea Musiu**   **20**   **Sudden death**   Jul 23, 2022
https://www.ilmessaggero.it/italia/andrea_musiu_morto_calcetto_partita_cagliari_chi_era_ultime_notizie-6831104.html

169 **Unnamed Person**   **20**   **Cardiac event during marathon**   May 22, 2022
https://www.waz.de/staedte/gelsenkirchen/zusammenbruch-vor-ziel-mann-stirbt-bei-vivawest-marathon-id235422969.html

170 **Oliver Vaux**   **20**   **Died unexpectedly in his sleep**   May 26, 2022
https://www.heraldscotland.com/news/20202278.university-pays-tribute-exemplary-student-oliver-vaux-dies-sleep/

171 **Andres Melendez**   **20**   **Sudden death**   Dec 16, 2021
https://www.cleveland.com/guardians/2022/01/what-caused-the-death-of-cleveland-guardians-minor-leaguer-andres-melendez-hey-hoynsie.html

172 **Joanna Krudys**   **21**   **Sudden death post vaccination**   Dec 4, 2021
https://nczas.com/2021/12/09/nagla-smierc-dwojki-wroclawskich-studentow-internet-wrze-foto/

173 **Awysum Harris**   **21**   **Sudden death**   Jul 3, 2022
https://www.waff.com/2022/07/05/alabama-state-mourns-death-football-player/

174 **Reda Saki**   **21**   **Sudden death during game**   Apr 11, 2021
https://www.moroccoworldnews.com/2021/04/339555/21-year-old-moroccan-football-player-dies-after-collapsing-on-pitch

175 **Alexey Zelenin**   **21**   **Cardiac event during training**   May 12, 2022
https://markcrispinmiller.substack.com/p/in-memory-of-those-who-died-suddenly-a59?s=r

176 **Fabricio Navarro**   **21**   **Died unexpectedly in his sleep**   Jun 15, 2022
https://442.perfil.com/noticias/futbol/dolor-en-atletico-tucuman-por-el-fallecimiento-de-uno-de-sus-jugadores.phtml

177 **Marvel Simiyu**   **21**   **Sudden death**   Jun 14, 2022
https://euroweeklynews.com/2022/06/16/young-female-footballer-marvel-simiyu-dies-suddenly/

178 **Aliya Khambikova**   **21**   **Sudden death**   Nov 7, 2021
https://www.rt.com/sport/539670-russian-volleyball-death-aliya-khambikova/

179 **Nelson Solano**   **21**   **Cardiac event**   Nov 8, 2021
https://www.abc.com.py/nacionales/2021/11/07/joven-futbolista-fallece-de-un-infarto-despues-de-un-partido/

180 **Dawid Akula**   **21**   **Sudden death during game**   Dec 4, 2021
https://nczas.com/2021/12/09/nagla-smierc-dwojki-wroclawskich-studentow-internet-wrze-foto/

181 **Aurelie Hans**   **21**   **Cardiac event**   Dec 14, 2021
https://www.dna.fr/societe/2021/12/18/apres-le-deces-d-aurelie-hans-l-emotion-dans-le-monde-du-football-feminin

182 **Bryce Murray**   **21**   **Sudden death**   Dec 27, 2021
https://www.thescottishsun.co.uk/news/scottish-news/8273835/forfar-scaffolder-suden-death-tributes/

183 **Herbert Afayo**   **21**   **Cardiac event during training**   Jan 2022
https://theinvestigatornews.com/2022/01/oh-no-the-sad-story-of-how-footballer-hebert-afayo-collapsed-and-instantly-died-on-pitch/

184 **Alexandros Lampis**   **21**   **Cardiac event during game**   Feb 2, 2022
https://www.thesun.co.uk/sport/17524888/alexandros-lampis-dies-21-greek-footballer-cardiac-arrest/

185 **Sanjay Vimalraj**   **21**   **Sudden death**   Jul, 2022
https://timesofindia.indiatimes.com/city/chennai/tamil-nadu-cm-announces-rs-3lakh-solatium-to-kin-of-deceased-kabaddi-player/articleshow/93166087.cms

186 **Clark Yarbrough**   **21**   **Sudden death**   Sep 4, 2022
https://www.si.com/college/2022/09/05/ouachita-baptist-defensive-lineman-clark-yarbrough-dies-at-21

187 **Renjitha**   **21**   **Blood clot days post vaccination**   Aug, 2021
https://keralakaumudi.com/en/news/news.php?id=625121&u=%20news.php?id=21-year-old-dies-in-kasargod-days-after-taking-covid-jab

188	**John Foley**	**21**	**Died in sleep hours post vaccination**	Apr 11, 2021

https://thecovidblog.com/2021/04/14/john-francis-foley-21-year-old-university-of-cincinnati-student-dead-24-hours-after-johnson-johnson-shot/

189	**Dominyka Podziute**	**21**	**Sudden death**	Apr 13, 2022

https://www.mirror.co.uk/sport/football/news/dominyka-podziute-dead-former-newcastle-26710872

190	**Laura Domecq**	**21**	**Sudden death**	Aug 7, 2022

https://www.monacomatin.mc/football/une-marche-organisee-ce-mercredi-soir-en-hommage-a-une-jeune-joueuse-de-las-monaco-decedee-791963

191	**Gianmarco Verdi**	**21**	**Sudden death during dinner**	May 28, 2022

https://www.ilrestodelcarlino.it/modena/cronaca/gianmarco-verdi-1.7726078

192	**Justin Tabone**	**22**	**Cardiac event during game**	Oct 10, 2022

https://timesofmalta.com/articles/view/22yearold-dies-collapsing-football-pitch.986852

193	**Hayden Holman**	**22**	**Cardiac event during marathon**	Oct 4, 2021

https://www.stgeorgeutah.com/news/archive/2022/09/29/ggg-friends-family-will-commemorate-hayden-holmans-life-at-46th-running-of-st-george-marathon/#.Y03wkC-B1-U

194	**Imtiyaz Khan**	**22**	**Cardiac event during game**	Sep 2, 2022

https://www.greaterkashmir.com/kashmir/pulwama-youth-dies-of-suspected-heart-attack-while-playing-cricket-in-anantnag

195	**Bruno Macedo**	**22**	**Sudden death**	Nov 21, 2021

https://www.ouest-france.fr/nouvelle-aquitaine/nueil-les-aubiers-79250/faits-divers-deces-d-un-jeune-homme-de-22-ans-il-jouait-au-fc-nueil-les-aubiers-122cc602-4af2-11ec-8a6b-582d17cbe42b

196	**Dejmi Dumervil**	**22**	**Sudden death**	Nov 12, 2021

https://sports.yahoo.com/former-louisville-football-player-dejimi-154847843.html

197	**Fatimah Shabazz**	**22**	**Sudden death**	Nov 30, 2021

https://greensboro.com/sports/college/a-t-volleyball-player-fatimah-shabazz-dies-suddenly/article_8c531018-521c-11ec-9cde-fb75a01ce59d.html

198	**Patricio Guaita**	**22**	**Sudden death during training**	Jan 18, 2022

https://www.ole.com.ar/informacion-general/fallecimiento-patricio-guaita-jugador-22-anos-plata_0_jvyHHTY7b.html

199	**Michael Almanza**	**22**	**Cardiac event during game**	Mar 20, 2022

https://euroweeklynews.com/2022/03/22/young-footballer-heart-attack/

200	**Arthur Grice**	**22**	**Neurological events day post vaccination**	Feb 24, 2022

https://circleofmamas.com/health-news/22-year-old-arthur-grice-died-6-weeks-after-johnson-johnson-vaccine-from-paralytic-ileus/

201	**Mubarak Sayed**	**22**	**Cardiac event**	Jul, 2022

https://www.mirror.co.uk/news/world-news/university-student-dies-joy-after-27613873

202	**Costa Debochado**	**22**	**Sudden death**	Jan 22, 2021

https://www.breaktudo.com/tiktoker-leo-costa-debochado-morre-aos-18-anos-de-idade-apos-passar-por-problemas-pulmonares/

203	**Maxime Beltra**	**22**	**Sudden death 9 hours post vaccination**	Jul 26, 2021

https://thecovidblog.com/2021/08/03/maxime-beltra-22-year-old-french-man-dead-nine-hours-after-first-experimental-pfizer-mrna-injection/

204	**Davis Heller**	**22**	**Sudden death**	Oct 6, 2022

https://www.usatoday.com/story/sports/2022/10/07/alabama-transfer-ngu-baseball-player-davis-heller-dies-at-age-22-north-greenville/69546077007/

205	**Andres Burgos**	**22**	**Cardiac event**	Jun 27, 2022
	https://www.molinabasket.es/es/publication/126756			

206	**Georgia Solanaki**	**22**	**Cardiac event during training**	Jun 15, 2022
	https://www.newsy-today.com/tragedy-during-training-the-22-year-old-died/			

207	**Samuel Carletti**	**22**	**Cardiac event**	Mar 22, 2022
	https://www.trevisotoday.it/cronaca/selva-volpago-montello-samuel-carletti-22-marzo-2022.html			

208	**Unnamed Person**	**22**	**Cardiac event during game**	Mar 23, 2022
	https://www.telenoche.com.uy/sociedad/fallecio-un-infarto-jugador-amateur-division-40-n5326902			

209	**Ben Penrose**	**22**	**Cardiac event**	Aug 16, 2022
	https://www.theleader.com.au/story/7870298/fundraiser-for-young-man-with-a-full-heart/			

210	**Rea Gostima**	**22**	**Died unexpectedly in her sleep**	Aug, 2022
	https://www.imolaoggi.it/2022/08/10/malore-improvviso-ragazza-di-22-anni-muore-nel-sonno/			

211	**Mattia Ghiraldi**	**22**	**Died unexpectedly in his sleep**	Jul, 2022
	https://primacremona.it/cronaca/mattia-ghiraldi-promessa-della-motonautica-scompare-improvvisamente-a-22-anni/			

212	**James Théodore**	**22**	**Cardiac event during game**	Mar 1, 2022
	https://www.eurosport.it/rugby/tragedia-in-francia-muore-in-allenamento-james-theodore-pilone-di-22-anni_sto8823863/story.shtml			

213	**Caitlin Gotze**	**23**	**Adverse events upon vaccination**	Nov 17, 2021
	https://www.covidvaccineinjuries.com/covid-vaccine-stories/caitlin-gotze-23-year-old-died-while-at-work-after-employer-mandated-covid-vaccine/			

214	**Antonio Salerno**	**23**	**Sudden death after game**	Sep 12, 2022
	https://www.ilrestodelcarlino.it/ferrara/cronaca/il-suo-sorriso-spento-per-sempre-in-un-attimo-e-stato-cancellato-tutto-1.8081608			

215	**Abdel Rahman**	**23**	**Sudden death during game**	Aug 3, 2021
	https://sportsbeezer.com/allsports/look-an-egyptian-player-who-swallowed-his-tongue-and-died/			

216	**Roy Butler**	**23**	**Sudden death 4 days post vaccination**	Aug 13, 2021
	https://thecovidblog.com/2021/08/23/roy-butler-23-year-old-irish-soccer-football-player-suffers-massive-brain-bleed-dead-four-days-after-experimental-johnson-johnson-viral-vector-dna-injection/			

217	**Gilbert Kwemoi**	**23**	**Sudden death**	Aug 14, 2021
	https://www.insidethegames.biz/articles/1111716/gilbert-soet-kwemoi-dies-aged-23			

218	**Riuler de Oliveira**	**23**	**Cardiac event**	Nov 23, 2021
	https://ge.globo.com/google/amp/pr/futebol/noticia/ex-athletico-e-coritiba-riuler-oliveira-morre-vitima-de-infarto-aos-23-anos.ghtml			

219	**Erik Karlsson**	**23**	**Cardiac event**	Dec 31, 2021
	https://www.aftonbladet.se/sportbladet/a/MLqj85/elitlopare-dod—fick-hjartstopp-under-lopp-i-kalmar			

220	**Branson King**	**23**	**Sudden death**	Dec 11, 2021
	https://www.hollywoodlanews.com/young-ice-hockey-player-dies-unexpectedly/			

221	**Marin Cacic**	**23**	**Cardiac event**	Dec 21, 2021
	https://www.index.hr/sport/clanak/mladi-nogometas-nehaja-iz-senja-se-srusio-na-treningu-bore-mu-se-za-zivot/2327080.aspx			

222 **Thottyanda Somanna   23   Cardiac event during game**   Dec 25, 2021
https://www.newindianexpress.com/states/karnataka/2021/dec/25/hockey-player-dies-
in-the-middle-of-a-game-inkarnataka-heart-attack-suspected-2399669.html

223 **Jamie Hoye   23   Sudden death**   Jan 9, 2022
https://www.armaghi.com/news/lurgan-news/tributes-paid-to-young-lurgan-man-jamie-
hoye-who-was-the-kindest-wee-soul/154422

224 **Michel Corbalan   23   Sudden death**   Jan 28, 2022
https://ekstrabladet.dk/sport/anden_sport/anden_sport/dansk-europamester-doed-23-
aar-gammel/9109434?ilc=c

225 **Mary Cronin   23   Sudden death**   Apr 29, 2022
https://www.cbc.ca/news/canada/new-brunswick/mary-cronin-fredericton-soccer-unb-
st-thomas-covid-19-1.6438272

226 **Michael Morgan   23   Sudden death**   Jul 4, 2022
https://www.belfastlive.co.uk/news/northern-ireland/coalisland-community-stunned-
after-sudden-24394381

227 **Amelia Smith   23   Died unexpectedly in his sleep**   Nov 14, 2021
https://www.dailymail.co.uk/news/article-10222507/Beautiful-kind-mother-two-23-dies-
suddenly-sleep-month-giving-birth.html

228 **Krzysztof Pańka   23   Sudden death**   Nov 28, 2021
https://sport.wprost.pl/10555684/krzysztof-panka-nie-zyje-klub-potwierdzil-informacje-
o-smierci-23-letniego-reprezentanta-polski.html

229 **Abraham Sié   23   Sudden death**   Apr 6, 2022
https://www.sport-ivoire.ci/basketball/décès-dabraham-sié-la-bal-attristée

230 **Hannah Langer   23   Died unexpectedly in her sleep**   Aug 27, 2022
https://highlandscurrent.org/2022/08/29/hannah-langer-1998-2022/

231 **Jessica Courtney   23   Sudden death**   Jan, 2022
https://worldnewsera.com/news/uk/woman-23-dies-suddenly-from-mystery-
undiagnosed-illness-as-family-pay-tribute/

232 **Traian Calancea   24   Stroke 10 days post vaccination**   Oct, 2021
https://lanuovabq.it/it/morto-a-24-anni-il-giudice-ordina-indagate-sul-vaccino

233 **Icaro Da Silva   24   Cardiac event**   Apr 30, 2021
https://portalmt.com.br/parada-cardiaca-morre-jogador-que-estava-em-periodo-de-
teste-na-equipe-do-acao-s-a-f-c/

234 **Josh Downie   24   Cardiac event**   May 10, 2021
https://www.bbc.com/news/uk-england-nottinghamshire-57058626

235 **Boris Sadecky   24   Cardiac event during game**   Nov 3, 2021
https://usdaynews.com/celebrities/celebrity-death/boris-sadecky-death-cause/

236 **Michal Krowiak   24   Sudden death**   Dec 1, 2021
https://wmeritum.pl/michal-krowiak-nie-zyje-pilkarz-mial-zaledwie-24-lata/363554

237 **Ahmed Amin   24   Cardiac event in locker room**   Dec 22, 2021
https://www.kingfut.com/2021/12/23/third-divisions-rabat-anwar-goalkeeper-dies-of-
cardiac-arrest/

238 **Dillon Quirke   24   Sudden death during game**   Aug 5, 2022
https://www.thetimes.co.uk/article/dillon-quirke-tipperary-hurler-had-spoken-of-his-
hearts-fatal-flaw-5r7k829kr

239 **Carla Steytler   24   Sudden death on campus**   Feb, 2022
https://www.thesouthafrican.com/news/student-who-collapsed-and-died-on-bloem-
campus-described-as-go-getter-breaking/

240	**Kim Lockwood**	24	**Vaccine-induced death, per coroner**	Mar 24, 2022
	https://www.bbc.com/news/uk-england-south-yorkshire-60757293			
241	**Joyce Culla**	24	**Brain aneurysm weeks after booster**	Apr 1, 2022
	https://community.covidvaccineinjuries.com/joyce-culla-24-year-old-nurse-and-tiktok-star-dies-after-ruptured-brain-aneurysm-following-astrazeneca-vaccine/			
242	**Elia Fiorio**	24	**Sudden death**	Aug 30, 2022
	https://euroweeklynews.com/2022/09/02/italian-man-dies-suddenly-unexpectedly-spain-mallorca/			
243	**Finley Scholefield**	24	**Died unexpectedly in his sleep**	Mar 21, 2022
	https://www.theguardian.com/technology/2022/apr/28/finley-scholefield-obituary			
244	**Matt Rodrigopulle**	24	**Sudden death**	Sep 4, 2022
	https://globalnews.ca/video/9118861/matthew-rodrigopulle-memorial/			
245	**Sam Polledri**	24	**Cardiac event**	Mar 2, 2022
	https://www.mirror.co.uk/sport/rugby-union/polledri-gloucester-italy-death-rugby-26366673			
246	**Unnamed Person**	25	**Sudden death during training**	Mar 2, 2022
	https://www.straitstimes.com/singapore/nsman-25-dies-after-collapsing-during-hpb-exercise-session			
247	**Giacomo Gorenszach**	25	**Sudden death**	Jun 2, 2022
	https://www.udinetoday.it/cronaca/morto-giacomo-gorenszach-san-pietro-natisone.html			
248	**Ashley Hipper**	25	**Sudden death**	Jan 12, 2022
	https://dailyvoice.com/new-jersey/sussex/obituaries/beloved-sussex-county-hs-grad-bio-technician-ashley-hipper-dies-suddenly-at-25/824599/			
249	**P Lerkchaleampote**	25	**Died unexpectedly in his sleep**	Mar 23, 2022
	https://www.straitstimes.com/life/entertainment/thai-actor-beam-papangkorn-lerkchaleampote-dies-suddenly-in-sleep-at-25			
250	**Fancis Perron**	25	**Sudden death after game**	Sep 18, 2021
	https://ottawasun.com/sports/football/tragedy-for-gee-gees-defensive-lineman-francis-perron-dies-after-game-in-toronto			
251	**Michelle de Vecchi**	25	**Cardiac event during run**	Nov 17, 2021
	https://corrieredelveneto.corriere.it/treviso/cronaca/21_novembre_17/treviso-muore-25-anni-facendo-jogging-un-amico-5e8e8e8c-47ef-11ec-a5cc-cbd997036243.shtml			
252	**N Mirosavljevic**	25	**Cardiac event**	Dec 24, 2021
	https://sportal.blic.rs/fudbal/domaci-fudbal/preminuo-nemanja-mirosavljevic-od-posledica-srcanog-udara/2022041920343298399			
253	**Marcos Menaldo**	25	**Cardiac event during training**	Jan 3, 2022
	https://www.dailystar.co.uk/sport/football/marcos-menaldo-footballer-dies-25-25847436			
254	**Abel Wasan**	25	**Found dead day post vaccination**	Aug 1, 2021
	https://rupreparing.com/news/2021/11/25/abel-wasam-25-year-old-programmer-dies-1-day-after-receiving-the-astrazeneca-covid-19-vaccine-family-seeks-answers			
255	**Desiree Penrod**	25	**Sudden death days post vaccination**	Mar 17, 2021
	https://thecovidblog.com/2021/03/22/desiree-penrod-25-year-old-connecticut-educator-dead-one-week-after-johnson-johnson-viral-vector-shot/			
256	**Olivia Quan**	25	**Brain bleed**	Jul 1, 2022
	https://euroweeklynews.com/2022/07/10/cause-death-recording-engineer-olivia-quan-died-suddenly/			
257	**Sofia Constantino**	25	**Sudden death**	Jan 27, 2022
	https://www.direttasicilia.it/2022/01/27/sofia-costantino-termini-imerese-tragedia/			

258 **Jordan Fitzgerald** **25** **Sudden death** Jul, 2022
https://www.thesun.ie/sport/9055587/jordan-fitzgerald-limerick-fan-died-croke-park-bus/

259 **Michele Gironella** **25** **Cardiac event during game** Aug 17, 2022
https://www.ilrestodelcarlino.it/macerata/cronaca/michele-gironella-1.7994646

260 **Caleb Swanigan** **25** **Sudden death** Jun 20, 2022
https://ftw.usatoday.com/lists/caleb-swanigan-death-25-years-old-purdue-portland

261 **Debajyoti Ghosh** **25** **Cardiac event during game** Mar 19, 2022
https://www.news9live.com/sports/football/shades-of-cristiano-junior-as-east-bengal-bound-footballer-debojyoti-ghosh-dies-on-field-160078

262 **Nathan Bellshaw** **25** **Died unexpectedly in his sleep** Apr 18, 2022
https://www.thecourier.co.uk/fp/news/dundee/3293057/nathan-bellshaw-death-dundee/

263 **Chi Bernard** **20s** **Sudden death** Apr 29, 2022
https://www.maravipost.com/nigerian-actress-chinedu-bernard-dies-while-cleaning-in-church/

264 **Eoghan Moloney** **20s** **Sudden death** Jun 17, 2022
https://www.thesun.ie/news/8972587/tributes-outstanding-gaa-player-dies-suddenly/

265 **Edier Armero** **20s** **Sudden death during game** Apr 4, 2022
https://noticias.caracoltv.com/ojo-de-la-noche/impactante-muerte-de-un-futbolista-se-desplomo-en-plena-cancha

266 **Augin Kasonga** **20s** **Cardiac event** Feb 2, 2022
https://ufc.divisionafrica.org/fr/services-aux-joueurs/us-tshinkunku-augustin-kasonga-meurt-dun-arret-cardiaque-2283

267 **Iago dos Santos** **26** **Cardiac event** Feb 27, 2022
https://gauchazh.clicrbs.com.br/esportes/noticia/2022/03/jogador-de-futebol-brasileiro-morre-na-franca-e-familia-faz-vaquinha-para-trazer-o-corpo-ao-brasil-cl09x7hmc00600165gzp7i358.html

268 **Anele Gasa** **26** **Sudden death during game** Jun 16, 2022
https://www.citizen.co.za/witness/news/durban/amateur-football-player-dies-during-a-match-20220617/

269 **Kevin Revillod** **26** **Cardiac event during training** Aug 16, 2022
https://www.sudinfo.be/id482885/article/2022-08-17/le-foot-provincial-en-deuil-kevin-26-ans-et-papa-dun-enfant-de-18-mois-seffondre

270 **Gillen Lusson** **26** **Died unexpectedly in his sleep** Dec 25, 2021
https://www.sportbusinessmag.com/handball/hand-carnet-noir-lancien-capitaine-du-sco-dangers-est-decede-le-jour-de-noel/

271 **Alan Mellouet** **26** **Sudden death during training** Jan 26, 2022
https://www.lelibrepenseur.org/alan-mellouet-footballeur-de-26-ans-meurt-apres-un-malaise-a-lentrainement/

272 **Ryan Buyting** **26** **Sudden death** Jul 26, 2022
https://yorkfh.com/tribute/details/7627/Ryan-Buyting/obituary.html

273 **Will Jones** **26** **Died unexpectedly in his sleep** Jun 2, 2022
https://www.nationalworld.com/news/uk/will-jones-sudden-death-dad-falling-asleep-family-devastated-3736174

274 **Nikita Lomax** **26** **Sudden death while napping** May, 2022
https://www.mirror.co.uk/news/uk-news/mum-26-tragically-dies-after-27116484

275	**Brian Wallace**	**26**	**Cardiac event**	Apr 15, 2022

https://razorbackswire.usatoday.com/2022/04/15/former-arkansas-lineman-brian-wallace-passes-away/

276	**Rory Nairn**	**26**	**Vaccine-induced death per coroner**	Nov 17, 2021

https://www.rnz.co.nz/news/national/475109/man-s-death-ruled-a-result-of-rare-reaction-to-covid-19-vaccine

277	**Andrea Dorno**	**26**	**Died unexpectedly in her sleep**	Mar 22, 2022

https://www.romatoday.it/zone/nomentano/san-lorenzo/morte-andrea-dorno-san-lorenzo.html

278	**Kaila Chizer**	**26**	**Sudden death**	Aug 23, 2022

https://tucson.com/sports/arizonawildcats/basketball/former-arizona-wildcats-support-staffer-kaila-chizer-dies-at-26/article_fe53aada-24d8-11ed-96fc-bbacb9ea2fe7.html

279	**Alessandro Tedde**	**26**	**Died unexpectedly in his sleep**	Apr, 2022

https://www.unionesarda.it/news-sardegna/cagliari/alessandro-tedde-stroncato-a-26-anni-da-un-malore-grande-cordoglio-da-cagliari-a-guamaggiore-jsgnbczp

280	**Erin Julia Beebe**	**26**	**Died unexpectedly in her sleep**	Apr 4, 2022

https://dailyvoice.com/new-jersey/rutherford/obituaries/beloved-sparta-high-school-grad-johns-hopkins-research-technologist-dies-suddenly-26/829833/

281	**Dr. Candace Nayman**	**27**	**Sudden death after triathlon**	Jul 28, 2022

https://torontosun.com/news/local-news/warmington-triathlete-27-becomes-5th-gta-doctor-to-die-in-july

282	**Sri Vishnu**	**27**	**Sudden death during workout**	Jun 4, 2022

https://www.thenewsminute.com/article/27-year-old-man-collapsesdies-after-gym-workout-tamil-nadu-164869

283	**Vishnuvardhan Reddy**	**27**	**Cardiac event while cycling**	Jul 2, 2022

https://thenewsglory.com/sad-exercise-young-techie-died-got-married-3-months-ago/

284	**Mariasofia Paparo**	**27**	**Cardiac event**	Apr 11, 2022

https://www.swim4lifemagazine.it/2022/04/11/nuoto-master-in-lutto-si-spegne-maria-sofia-paparo/

285	**Joseph Wandera**	**27**	**Sudden death**	Jan, 2022

https://metro.co.uk/2022/02/24/coronation-street-actor-died-of-sudden-illness-while-abroad-16166870/

286	**Niels De Wolff**	**27**	**Cardiac event after game**	Mar 10, 2021

https://www.hln.be/waasmunster/voetbalspeler-niels-de-wolf-27-overleden-nadat-hij-zondag-werd-getroffen-door-hartfalen-na-wedstrijd~abb588b5/

287	**Haziq Kamaruddin**	**27**	**Sudden death 8 days post vaccination**	May 14, 2021

https://www.scmp.com/sport/other-sport/article/3133669/haziq-kamaruddin-malaysia-mourns-death-olympic-archer-27-health

288	**Yusuke Kinoshita**	**27**	**Cardiac event 8 days post vaccination**	Aug 3, 2021

https://www.nikkansports.com/baseball/news/202108060001163.html

289	**Tim B.**	**27**	**Sudden death**	Jul 23, 2021

https://www.tag24.de/nachrichten/regionales/schleswig-holstein/fussballer-gestorben-https-www-instagram-com-p-crevyz0gepa-utm-medium-copy-link-2054238

290	**Yvonne Jelagat Morwa**	**27**	**Sudden death**	Sep 6, 2021

https://athletics.co.ke/shock-as-another-kenyan-athlete-passes-on/

291	**Hector Vilellas Soro**	**27**	**Cardiac event during race**	Nov 18, 2021

https://euroweeklynews.com/2021/11/18/young-athlete-dies-after-competing-in-behobia-san-sebastian-event/

292 **Endy Maldonado** **27** **Cardiac event during game** Jan 20, 2022
https://hoy.com.do/fallece-nieto-de-ruben-maldonado-mientras-jugaba-baloncesto/

293 **Tyler Edwards** **27** **Cardiac event** Aug 15, 2022
https://www.woodtv.com/news/kent-county/comstock-park-hs-varsity-boys-basketball-coach-dies-at-27/

294 **Vanessa Lisitano** **27** **Sudden death** Jan 15, 2022
https://www.ilroma.net/news/attualità/maestra-trovata-morta-il-suo-amico-dario-commosso-era-una-ragazza-brillante-non-ci

295 **Jack Last** **27** **Blood clot 21 days post vaccination** Apr 20, 2021
https://www.dailymail.co.uk/news/article-9516013/Fit-healthy-engineer-27-died-three-weeks-having-AstraZeneca-anti-Covid-jab.html

296 **Dr. Joshimar Henry** **27** **Cardiac event** Apr 3, 2021
https://circleofmamas.com/health-news/vaccinated-dr-joshimar-henry-dies-at-27-of-heart-attack/

297 **William Harding** **27** **Sudden death** Jun 25, 2022
https://www.voiceofalexandria.com/announcements/funeral/obituary—william-m-will-harding-27/article_8ee57cac-f88f-11ec-8976-534d7348e470.html

298 **Anna Kruglova** **27** **Cardiac event 2 days post vaccination** Sep 30, 2021
https://goodsciencing.com/covid/athletes-suffer-cardiac-arrest-die-after-covid-shot/

299 **Diego Romito** **27** **Died unexpectedly in his sleep** Jul, 2022
https://www.veronaoggi.it/cronaca/tragedia-notte-studente-muore-sonno-pressana-12-luglio-2022/

300 **Matteo Lorenzato** **27** **Died unexpectedly in his sleep** Apr, 2022
https://www.imolaoggi.it/2022/04/29/malore-nel-sonno-matteo-lorenzato-muore-a-27-anni/

301 **Scott MacDonald** **27** **Cardiac event while making dinner** Feb 23, 2022
https://www.thesun.ie/news/8449509/tributes-scott-murray-macdonald-irish-fitness-coach-death/

302 **Niall Sammon** **27** **Sudden death** Jan 1, 2022
https://www.sundayworld.com/news/irish-news/young-man-who-died-suddenly-on-new-years-day-described-as-gentle-giant/41205680.html

303 **Ulfrido García** **28** **Cardiac event** Mar 4, 2022
http://www.5septiembre.cu/muere-lanzador-santiaguero-ulfrido-garcia/

304 **John Paul** **28** **Died unexpectedly in his sleep** Mar 9, 2022
https://www.mirror.co.uk/sport/other-sports/cycling/john-paul-scotland-cyclist-dies-26430280?utm_source=twitter.com&utm_medium=social&utm_campaign=sharebar

305 **Dimitri Roveri** **28** **Cardiac event during game** Apr 7, 2022
https://milano.corriere.it/notizie/lombardia/22_maggio_09/dimitri-roveri-futuro-papa-morto-giocando-calcio-eri-nostro-capitano-aiuteremo-tuo-figlio-37f9e388-cf09-11ec-b21c-2cc92198dafc.shtml

306 **Kalindi Souza** **28** **Cardiac event** Apr 10, 2022
https://www.world-today-news.com/kalindi-ex-academica-and-nacional-died-at-the-age-of-28/

307 **Wade Badara** **28** **Cardiac event during training** Dec 30, 2021
http://www.fajarinfos.com/2021/12/31/un-joueur-du-duc-meurt-a-la-suite-dun-malaise/

308 **Kamil Pulczynski** **28** **Cardiac event** Jan 22, 2022
https://sportowefakty.wp.pl/zuzel/981308/mial-zaledwie-28-lat-polozyl-sie-spac-jak-zwykle-ale-juz-nigdy-sie-nie-obudzil

309 **Mohammad Rashid** **28** **Cardiac event during workout** Jan 18, 2022
https://kashmir.today/sopore-mbbs-student-dies-of-heart-attack/

310 **David Tong Tjouen** **28** **Sudden death** Jan 30, 2022
https://actu.fr/ile-de-france/sarcelles_95585/sarcelles-le-karateka-tong-tjouen-est-decede-a-28-ans_48519189.html

311 **Eric Robertson** **28** **Sudden death** Mar 7, 2022
https://www.knoxnews.com/story/sports/2022/03/08/knoxville-teacher-basketball-coach-eric-robertson-dies-suddenly/9428393002/

312 **Akhlad Khan** **28** **Cardiac event** Apr 11, 2022
https://community.covidvaccineinjuries.com/akhlad-khan-28-year-old-times-of-india-journalist-has-died-after-3-heart-attacks/

313 **Sara Stickles** **28** **Stroke 5 days post vaccination** Feb 17, 2021
https://circleofmamas.com/health-news/28-year-old-mother-has-stroke-5-days-after-covid-vaccine/

314 **Antonio Frangiosa** **28** **Sudden death** May, 2022
https://www.imolaoggi.it/2022/05/16/muore-improvvisamente-antonio-frangiosa/

315 **Haley Brinkmeyer** **28** **Sudden death 2 days post vaccination** Jan 21, 2021
https://www.dailymail.co.uk/news/article-9357521/Physical-therapist-28-dies-two-days-taking-COVID-19-vaccine.html

316 **Dominic Green** **28** **Cardiac event** Jan 17, 2022
https://www.msn.com/en-us/news/us/he-worked-from-home-and-died-suddenly-five-days-passed-before-his-body-was-found/ar-AAZumbl

317 **Saron Berhe** **28** **Sudden death** Jan 17, 2022
https://www.12news.com/article/news/community/accomplished-valley-woman-dies-unexpectedly-while-in-pursuit-of-law-degree/75-5e34733c-975b-4617-af91-8c117669ede9

318 **Cédric Baekeland** **28** **Cardiac event** Mar 14, 2022
https://blazetrends.com/belgian-cyclist-cedric-baekeland-dies-at-28-after-suffering-a-heart-attack-in-mallorca/

319 **Marcus Feeney** **28** **Died unexpectedly in his sleep** Sep 1, 2022
https://tinyurl.com/v93a6zfx

320 **Ojima Omale** **28** **Died unexpectedly in his sleep** Feb, 2022
https://thewhistler.ng/osinbajo-grieves-as-his-shoemaker-friend-dies-suddenly/

321 **Lee Moses** **29** **Sudden death during training** Aug 12, 2021
https://www.stuff.co.nz/manawatu-standard/news/300387209/young-father-and-footballer-dies-of-heart-attack-during-training

322 **Alexsandr Kozlov** **29** **Blood clot during training** Jul 15, 2022
https://euroweeklynews.com/2022/07/15/breaking-news-russian-footballer-aleksandr-kozlov-dies-after-blood-clot-aged-29/

323 **Giuseppe Perrino** **29** **Cardiac event** Jun 4, 2021
https://www.thesun.co.uk/sport/football/15151824/giuseppe-perrino-dead-29-parma-brother/

324 **Sebastian Eubank** **29** **Cardiac event** Jul 13, 2021
https://www.bbc.com/news/uk-57828270

325 **Arthur Zuccolini** **29** **Died unexpectedly in his sleep** Jul 15, 2021
https://www.leprogres.fr/societe/2021/07/15/l-ancien-choralien-arthur-zuccolini-n-est-plus

326 **Avi Barot** **29** **Cardiac event** Sep 16, 2021
https://www.thehindu.com/sport/cricket/young-saurashtra-cricket-player-avi-barot-dies-after-suffering-cardiac-arrest/article37015873.ece

327 **Dave Hyde** **29** **Sudden death during game** Sep 4, 2021
https://www.mirror.co.uk/news/uk-news/gentle-giant-dad-29-collapses-24933421

328 **Riccardo Firrarello** **29** **Sudden death** Oct 10, 2021
https://primabiella.it/cronaca/addio-a-riccardo-firrarello-papa-di-soli-29-anni/

329 **Daire Ni Heidhin** **29** **Sudden death** Nov 27, 2021
https://www.thesun.co.uk/news/16916240/tributes-paid-dundalk-woman-daire-ni-heidhin-died-prague/

330 **Mukhaled Al-Raqadi** **29** **Cardiac event during warm up** Dec 21, 2021
https://www.the-sun.com/sport/football/premier-league/4327481/oman-international-dies-collapsing-warm-up/

331 **Dani Chabrera** **29** **Sudden death** Dec 16, 2021
https://golsmedia.com/comunidad-valenciana/futbol/villarreal-cf/2021/12/16/fallece-dani-chabrera-preparador-porteros-exjugador-villarreal-cf/

332 **Kerim Arslan** **29** **Cardiac event after training** Jan 16, 2022
https://www.bunte.de/panorama/news-aus-aller-welt/kerim-arslan-29-fussballer-bricht-stunden-nach-training-zusammen-und-stirbt.html

333 **Ruben Michel** **29** **Cardiac event** Jan 16, 2022
https://www.patosonline.com/professor-de-educacao-fisica-morre-de-infarto-aos-29-anos-jovem-era-filho-de-casal-bastante-conhecido-em-patos/

334 **Gabriela Lessa** **29** **Cardiac event** Jan 27, 2022
https://twitter.com/Vivian2022N/status/1552239312263151618

335 **Cameron Dale** **29** **Stroke** Aug 30, 2021
https://www.dailymail.co.uk/news/article-9949391/Jessica-Watson-boyfriend-Cameron-Dale-died-stroke-age-29-Hamilton-Island-Queensland.html

336 **Manisha Yadav** **29** **Stroke** Oct 1, 2021
https://indianexpress.com/article/entertainment/television/jodha-akbar-actor-manisha-yadav-dies-7548842/

337 **Claudia Calcada** **29** **Cardiac event during marathon** Apr 30, 2022
https://cnnportugal.iol.pt/claudia-calcada/boavista/claudia-calcada-antiga-jogadora-do-boavista-morre-aos-29-anos

338 **Yennifer Ramirez** **29** **Cardiac event** Mar 23, 2022
https://www.diariolibre.com/deportes/deportes/2022/03/23/fallece-la-hija-de-soterio-ramirez/1725792

339 **Michele Lo Conte** **29** **Died unexpectedly in his sleep** Dec 29, 2021
https://www.calcionapoli24.it/notizie/ariano-che-choc-michele-muore-nel-sonno-a-29-anni-n505312.html

340 **Paddy Branagan** **30** **Sudden death** May 15, 2022
https://www.independent.ie/irish-news/news/talented-laois-hurler-dies-suddenly-in-us-41654798.html

341 **Jajov Adenan** **30** **Died at game days post vaccination** Nov 7, 2021
https://www.iltempo.it/attualita/2021/11/12/news/morto-calcetto-tre-giorni-dopo-seconda-dose-vaccino-covid-autopsia-sano-viterbo-covid-29418914/

342 **Jose Edgar Preciado** **30** **Cardiac event** Jun 19, 2021
https://www.thecaddienetwork.com/caddie-alberto-olguin-collapses-dies-during-pga-tour-latinoamerica-event/

343 **Ben Benn** 30 **Sudden death** Aug 22, 2022
https://newspunch.com/family-friends-in-shock-at-sudden-death-of-30-yr-old-rugby-player/

344 **Sofiane Loukar** 30 **Cardiac event after game** Dec 25, 2021
https://www.gulftoday.ae/sport/2021/12/25/algerian-football-player-sofiane-lokar-dies-of-heart-attack-during-match

345 **Maxim Ishkeldin** 30 **Blood clot in sleep** Jun 5, 2021
https://www.tellerreport.com/sports/2021-06-07-"it's-impossible-to-believe-such-news-right-away"—world-field-hockey-champion-maxim-ishkeldin-dies-at-the-age-of-30.rJMZiQwj5d.html

346 **Dimitri Ilongo** 30 **Sudden death** Jul 6, 2021
https://www.lechorepublicain.fr/chartres-28000/sports/disparition-de-l-ancien-coureur-de-chartres-dimitri-ilongo_13981279/

347 **Alexaida Guedez** 30 **Cardiac event** Aug 21, 2021
https://www.elnacional.com/deportes/atleta-alexaida-guedez-murio-durante-una-carrera-en-naguanagua/

348 **Remigio Bova** 30 **Died unexpectedly in her sleep** Oct 22, 2021
https://www.napolitoday.it/cronaca/morto-remigio-bova.html

349 **Muhammad Islam** 30 **Cardiac event during game** Oct 28, 2021
https://www.gurualpha.com/news/players-die-of-heart-attack-during-football-match/

350 **Aleksandar Krsic** 30 **Cardiac event** Nov 17, 2021
https://www.b92.net/sport/fudbal/vesti.php?yyyy=2021&mm=11&dd=19&nav_id=2059126

351 **Navid Khosh Hava** 30 **Cardiac event** Dec 5, 2021
https://www.tehrantimes.com/news/467718/Former-Iran-U23-defender-Navid-Khosh-Hava-dies

352 **Oisin Fields** 30 **Sudden death during game** Jan 5, 2022
https://www.armaghi.com/news/armagh-news/armagh-man-oisin-fields-had-a-huge-heart-who-spread-so-much-fun-and-generosity-in-this-world/154313

353 **Munther Al-Harassi** 30 **Cardiac event** Jan 23, 2022
https://www.tellerreport.com/sports/2022-01-24-omani-player-munther-al-harrasi-dies-on-the-field.ByIMxMU2TF.html

354 **German Clop** 30 **Died unexpectedly in his sleep** Jan 23, 2022
https://www.losandes.com.ar/mas-deportes/consternacion-en-el-futbol-mendocino-por-la-muerte-de-german-clop-ex-jugador-y-preparador-fisico/

355 **Sara Lee** 30 **Sudden death** Oct 6, 2022
https://www.tmz.com/2022/10/06/wwe-tough-enough-winner-sara-lee-dead-at-30/

356 **Jack Craig** 30 **Died unexpectedly in his sleep** Mar 17, 2022
https://www.msn.com/en-gb/health/mindandbody/happy-and-passionate-north-london-man-30-died-suddenly-in-his-sleep-despite-treatable-heart-condition/ar-AAVyiZI

357 **Ashley Gerren** 30 **Sudden death** Apr 12, 2021
https://nypost.com/2021/04/12/baldwin-hills-star-gerren-taylor-dead-at-30/

358 **Kevin Gregory** 31 **Cardiac event** Feb 15, 2022
https://www.irishnews.com/news/northernirelandnews/2022/02/16/news/co-down-man-31-dies-suddenly-in-ukraine-2589864/

359 **L McCutcheon** 31 **Sudden death** Sep, 2022
https://www.news.com.au/entertainment/celebrity-life/love-actually-star-martine-mccutcheons-brother-dies-aged-31/news-story/f87669063fe0413b9dd7323f820a642b

COMPENDIUM SAMPLING

360 Osahon Osawe   31   Cardiac event during game   May 26, 2022
https://www.leicestermercury.co.uk/news/local-news/young-dad-dies-playing-football-7278799

361 Mario Cunha   31   Sudden death   Aug 22, 2022
https://www.world-today-news.com/varzims-former-player-mario-cunha-has-died-aged-31/

362 Jimmy Hayes   31   Sudden death   Aug 23, 2021
https://www.bostonglobe.com/2021/08/23/sports/jimmy-hayes-former-bruin-boston-college-hockey-champ-dies-31/

363 Alex Evans   31   Cardiac event   Aug 21, 2021
https://www.bbc.com/news/uk-wales-58297010

364 Guillermo Arias   31   Sudden death during game   Sep 25, 2021
https://eltubazodigital.com/sucesos/guillermo-arias-fallece-deportista-del-camaguan-fc-en-plena-cancha-de-juego/2021/09/27/

365 David Jenkins   31   Sudden death   Oct 14, 2021
https://www.insidethegames.biz/articles/1114187/diving-coach-david-jenkins-olympics

366 Katie Novak   31   Sudden death   Jan 18, 2022
https://www.kare11.com/article/news/local/community-rallies-behind-family-of-popular-trainer-mother-who-died-unexpectedly/89-294c8106-8d5a-4651-969a-9360d59031b2

367 Mahima Matthew   31   Stroke days post vaccination   Aug 6, 2021
https://keralakaumudi.com/en/news/news-amp.php?id=622988&u

368 Kostas Mastrovasilis   31   Blood clot 20 days post vaccination   Sep 4, 2021
https://healthimpactnews.com/2021/27247-deaths-2563768-injuries-following-covid-shots-in-european-database-taiwan-records-more-deaths-from-vaccine-than-virus/

369 Chloe Haworth   31   Sudden death   Apr, 2022
https://www.edinburghlive.co.uk/news/edinburgh-news/edinburgh-mum-devastated-after-sudden-23837565

370 Catherine Keane   31   Died unexpectedly in her sleep   Jul 6, 2021
https://www.thesun.co.uk/health/18733216/fit-healthy-daughter-died-suddenly-sleep/

371 Andrea Murtas   31   Sudden death   May 24, 2022
https://www.lastampa.it/torino/2022/05/25/news/morto_improvvisamente_ad_almese_il_rugbista_andrea_murtas-4711180/

372 Dean Chiazari   31   Cardiac event during canoe race   Oct 8, 2021
https://www.algoafm.co.za/local/post-mortems-to-be-conducted-on-paddlers-who-died-at-fish-river-canoe-marathon

373 Mattia Van Glabbeek   31   Died unexpectedly in his sleep   May 22, 2022
https://www.imolaoggi.it/2022/05/26/prato-mattia-van-glabbeek-muore-nel-sonno-a-31-anni/

374 Levan Kbilashvili   31   Died unexpectedly in his sleep   Jun 21, 2022
https://muz-tv.ru/news/umer-uchastnik-shou-golos-levan-kbilashvili/

375 Barry Horn   31   Sudden death   Aug 30, 2022
https://www.glasgowlive.co.uk/news/airdrie-man-found-dead-spanish-25006144

376 Eshe Symonds   31   Sudden death   Jan 13, 2022
https://www.birminghammail.co.uk/black-country/mystery-surrounds-death-31-year-22818279

377 Mohd Sobri   31   Cardiac event   Jan 28, 2022
https://goodsciencing.com/covid/athletes-suffer-cardiac-arrest-die-after-covid-shot/

| 378 | Ilaria Pappa | 31 | Blood clot days post vaccination | Mar 20, 2021 |

https://www.ilgolfo24.it/sgomento-a-ischia-muore-docente-di-31-anni-al-cardarelli/

| 379 | Charlbi Dean | 32 | Sudden death | Aug 29, 2022 |

https://deadline.com/2022/08/charlbi-dean-dead-triangle-of-sadness-actor-was-32-obituary-1235103229/

| 380 | Adriano | 32 | Cardiac event during game | Jan 29, 2022 |

https://www.alagoas24horas.com.br/1414485/jovem-infarta-em-campo-de-futebol-e-chega-ao-hospital-sem-vida/

| 381 | Miloš Đorđević | 32 | Died unexpectedly in his sleep | Feb 11, 2022 |

https://www.telegraf.rs/sport/fudbal/3300715-futsal-smrt-milos-djordjevic

| 382 | Jeffrey Klumpkens | 32 | Sudden death | Feb 23, 2022 |

https://www.nieuwsblad.be/cnt/dmf20220223_92511275

| 383 | Thushar Bedwa | 32 | Cardiac event during game | Aug 9, 2022 |

https://telanganatoday.com/hyderabad-man-collapses-while-playing-cricket-dies-in-hospital

| 384 | Ria Vannoort | 32 | Sudden death while jet skiing | Aug 21, 2022 |

https://toronto.ctvnews.ca/ontario-paramedic-dies-suddenly-while-on-jet-ski-at-u-s-cottage-family-says-1.6039160

| 385 | Franciskao El Diex | 32 | Cardiac event during game | Dec 7, 2021 |

https://www.critica.com.pa/sucesos/muere-musico-panameno-de-paro-cardiaco-mientras-jugaba-futbol-618033

| 386 | Alfredo Quintana | 32 | Cardiac event | Feb 22, 2021 |

https://www.ihf.info/media-center/news/alfredo-quintana-warrior-who-left-too-early

| 387 | Tremaine Stewart | 32 | Sudden death during game | Apr 18, 2021 |

https://www.voice-online.co.uk/sport/football/2021/04/19/former-reggae-boy-tremaine-tan-tan-stewart-collapses-during-match-and-dies/

| 388 | Haitem Fathallah | 32 | Cardiac event during game | Oct 17, 2021 |

https://middleeast.in-24.com/sport/News/96094.html

| 389 | Stevan Jelovac | 32 | Stroke | Dec 5, 2021 |

https://www.eurohoops.net/en/fiba-champions-league/1274694/stevan-jelovac-dies-after-stroke/

| 390 | Monique Piske | 32 | Cardiac event | Jan 2, 2022 |

https://www.contrafatos.com.br/infarto-mata-campea-de-muay-thai-aos-32-anos/

| 391 | Benjamin Goodman | 32 | Cardiac event day post vaccination | Mar 14, 2021 |

https://thecovidblog.com/2021/03/21/benjamin-goodman-32-year-old-new-york-stagehand-dead-24-hours-after-johnson-johnson-viral-vector-shot/

| 392 | Nick Nemeroff | 32 | Died unexpectedly in his sleep | Jun 27, 2022 |

https://nypost.com/2022/06/28/comedian-nick-nemeroff-dead-at-32-fans-mourn-sudden-passing/

| 393 | Dr. Cason King | 32 | Sudden death | Aug 11, 2022 |

https://www.cressfuneralservice.com/obituary/Cason-KingPhD

| 394 | Dr. Stephen Wright | 32 | Blood clot post vaccination | Jan 26, 2021 |

https://expose-news.com/2021/06/08/nhs-doctor-died-suddenly-from-rare-stroke-due-to-blood-clots-after-receiving-astrazeneca-covid-19-vaccine/

| 395 | Dr. Stephanie Bosch | 32 | Blood clot | Oct 13, 2021 |

https://yachatsnews.com/waldport-clinic-doctor-stephanie-bosch-dies-suddenly-and-unexpectedly-wednesday-night-at-newport-hospital/

396	Brandon Pollet	33	Became sick 2 days post vaccination	Jan 28, 2022
	https://community.covidvaccineinjuries.com/brandon-pollet-33-year-old-husband-and-father-dies-after-developing-auto-immune-disease-hlh-after-pfizer-vaccine/			
397	Devaraj Anchan	33	Sudden death during game	Mar 29, 2021
	https://www.thehindu.com/news/cities/Mangalore/volleyball-player-collapses-during-game-dies/article34185430.ece			
398	Antron Pippen	33	Sudden death	Apr 18, 2021
	https://www.today.com/parents/scottie-pippen-opens-about-death-son-antron-t224262			
399	Gayan Shanaka	33	Cardiac event	Oct 10, 2021
	https://srilankamirror.com/sport/25185-sla-rugby-player-dies-of-heart-attack-during-training			
400	Ronald Mudimu	33	Stroke	Nov 28, 2021
	https://www.chronicle.co.zw/triangle-utd-player-dies/			
401	Daniel Ferguson	33	Sudden death	Dec 12, 2021
	https://www.belfastlive.co.uk/news/northern-ireland/daniel-ferguson-tributes-one-kind-22472500			
402	Charlotte Fisher	33	Sudden death	Jul 15, 2022
	https://www.nottinghampost.com/news/nottingham-news/amazing-mum-three-died-suddenly-7363063			
403	Jesse Boyden	33	Sudden death	Jul 8, 2022
	https://patch.com/new-jersey/newbrunswick/one-njs-best-wrestlers-dies-suddenly-age-33			
404	Arianna Mora	33	Died unexpectedly in her sleep	Aug, 2022
	https://tinyurl.com/2p9f6y8m			
405	Catalina Forero	33	Blood clot	Mar 24, 2022
	https://lfpress.com/news/local-news/young-colombian-mother-starting-new-life-in-canada-dies-suddenly-in-london			
406	Dr. Ben Campbell	33	Sudden death	Nov 2, 2021
	https://www.belfasttelegraph.co.uk/news/northern-ireland/tributes-paid-to-inspirational-co-antrim-psychologist-who-died-suddenly-41028833.html			
407	Aldiglade Bhamu	34	Sudden death	May 26, 2022
	https://www.newzimbabwe.com/mighty-warriors-star-bhamu-dies-aged-34/			
408	Danielle Hampson	34	Died unexpectedly in her sleep	Jun 18, 2022
	https://people.com/tv/tom-mann-still-in-shock-3-weeks-after-dani-hampson-death/			
409	Mitch Williams	34	Sudden death	Jul 2, 2022
	https://www.mirror.co.uk/news/world-news/young-teacher-suddenly-drops-dead-27409904			
410	Jorge Valencia	34	Cardiac event	Sep 4, 2022
	https://www.archyde.com/medellin-marathon-this-is-the-fatality-that-left-the-race-medellin-colombia/			
411	Jamie Harper	34	Sudden death	Sep 11, 2022
	https://www.birminghammail.co.uk/news/midlands-news/dad-be-34-dies-suddenly-25012158			
412	Dale Best	34	Sudden death during game	Feb 20, 2021
	https://www.brisbanetimes.com.au/national/queensland/rugby-league-player-remembered-as-amazing-bloke-after-dying-on-field-20210221-p574g0.html			
413	Leandro Siqueira	34	Cardiac event	Aug 17, 2021
	https://www.santaluzia-online.com/2021/08/jovem-atleta-do-futebol-tem-ataque.html			

414	**Zhora Khoroliev**	**34**	**Sudden death**	Dec 21, 2021

https://ligawl.pl/index.php/2021/12/29/michal-gurajdowski-nie-zyje/

415	**Filip Samardzic**	**34**	**Sudden death**	Jan 6, 2022

https://mondo.rs/Sport/Ostali-sportovi/a1579886/Preminuo-odbojkas-Filip-Samardzic.html

416	**Brandon Watt**	**34**	**Cardiac event / Myocarditis**	Nov 4, 2021

https://www.australiannationalreview.com/covid-19-deaths-and-injuries/brandon-watt-widow-of-34-year-old-canadian-man-admits-i-was-a-sheep-before-her-fully-vaxxinated-husband-dropped-dead-in-front-of-their-baby-girls/

417	**Salvatore Cadau**	**34**	**Cardiac event**	May, 2022

https://www.casedduonline.it/elmas-muore-a-34-anni-stroncato-da-un-infarto-parenti-e-amici-sotto-choc-per-salvatore-boris-cadau/

418	**Breck Denny**	**34**	**Sudden death**	Jan 17, 2022

https://news.yahoo.com/breck-denny-comedic-actor-writer-031500124.html

419	**Miranda Freyer**	**34**	**Died unexpectedly in her sleep**	Jan 6, 2022

https://www.the-sun.com/entertainment/4437296/miranda-fryer-cause-death-neighbours-heart-sleep/

420	**Stefano Russo**	**34**	**Cardiac event during race**	May 30, 2022

https://www.ilrestodelcarlino.it/cesena/cronaca/stefano-russo-spartan-race-1.7732053

421	**Sergio Satriano**	**35**	**Sudden death**	Feb 16, 2022

https://www.ilmattino.it/salerno/sergio_satriano_morto_calciatore_battipaglia-6508231.html

422	**Vinaya Vittal**	**35**	**Cardiac event during workout**	Mar 28, 2022

https://www.republicworld.com/india-news/city-news/bengaluru-35-year-old-woman-dies-due-to-heart-attack-while-working-out-in-gym-articleshow.html

423	**Chimena**	**35**	**Adverse events hrs post vaccination**	Aug 28, 2021

https://community.covidvaccineinjuries.com/chimena-35-years-old-and-suffered-two-strokes-6-days-after-the-pfizer-covid-vaccine/

424	**Robert Rosser**	**35**	**Cardiac event**	Nov 20, 2021

https://www.fox17online.com/homepage-showcase/he-loved-the-youth-of-grand-rapids-coach-rob-remembered-for-his-work-with-kids

425	**Anne VanGeest**	**35**	**Sudden death days post vaccination**	Apr 19, 2021

https://www.fox17online.com/news/local-news/ionia/ionia-co-womans-death-after-getting-j-j-vaccine-reported-to-the-cdc

426	**Toro Bill Jr**	**35**	**Cardiac event**	Apr 26, 2022

https://www.the-sun.com/sport/5206050/toro-bill-jr-dead-mexican-wrestler-heart-attack/

427	**Paul Duncan**	**35**	**Cardiac event**	Jul 15, 2022

https://www.dailymail.co.uk/news/article-11039535/Tragedy-former-Denver-Broncos-offensive-linesman-Notre-Dame-star-dies-aged-just-35.html

428	**Daniele Sansone**	**35**	**Sudden death**	Nov 29, 2021

https://citymilano.com/2021/12/01/il-calciatore-daniele-sansone-trovato-morto-in-casa-a-35-anni/

429	**Doudou Faye**	**35**	**Cardiac event**	Oct 30, 2021

https://new.in-24.com/sport/basketball/263667.html

430	**Victor Plakhuta**	**35**	**Sudden death days post vaccination**	Nov 6, 2021

https://ngs24-ru.turbopages.org/ngs24.ru/s/text/health/2021/11/09/70242362/

431	**Kane van Dierman**	**35**	**Sudden death**	Dec 12, 2021

https://www.mixedarticle.com/kane-van-diermen-death-cause-how-did-he-die/

432	**Suliman Mohamed**	35	**Sudden death**	Dec 19, 2021
	https://lecollimateur.ma/66385			
433	**Alfie Nunn**	35	**Cardiac event during game**	Jan 27, 2022
	https://www.eventiavversinews.it/e-morto-ieri-27-gennaio-per-attacco-cardiaco-durante-una-partita-a-dubai-il-calciatore-londinese-35enne-alfie-nunn/			
434	**Adam Strachan**	35	**Sudden death**	Jul 10, 2022
	https://www.dailyrecord.co.uk/news/scottish-news/tributes-paid-former-scots-footballer-27442768			
435	**Subrata Kakati**	35	**Cardiac event**	Jun 13, 2022
	https://www.time8.in/correspondent-of-private-satellite-channel-from-assams-lakhimpur-dies-suddenly-at-35/			
436	**Shehzad Azam**	36	**Cardiac event**	Sep 30, 2022
	https://www.latestly.com/socially/sports/cricket/shehzad-azam-rana-pakistans-36-year-old-first-class-cricketer-dies-of-cardiac-arrest-4270612.html			
437	**Unnamed Person**	36	**Cardiac event during bike race**	Aug 20, 2022
	https://www.archyde.com/evolene-vs-36-year-old-collapses-during-mountain-bike-races-dead/			
438	**Unnamed Person**	36	**Cardiac event during marathon**	Oct 2, 2022
	https://people.com/sports/london-marathon-runner-died-after-collapsing-during-race/			
439	**Maicol Orlandi**	36	**Sudden death**	Jun 25, 2022
	https://pledgetimes.com/mystery-about-the-death-of-maicol-found-lifeless-in-his-bed-at-the-age-of-36/amp/			
440	**Marquis Kido**	36	**Cardiac event during game**	Jun 14, 2021
	https://www.reuters.com/lifestyle/sports/indonesian-doubles-star-kido-dies-heart-attack-36-2021-06-14/			
441	**Rob Woodward**	36	**Sudden death**	Dec 6, 2021
	https://www.facebook.com/SJFC2016/posts/1488455288195593			
442	**Daniel Perkins**	36	**Cardiac event**	Dec 8, 2021
	https://7news.com.au/news/public-health/family-of-nsw-father-daniel-perkins-slams-politicians-disgraceful-covid-vaccine-claim-c-6273002			
443	**Catherine Fleagle**	36	**Sudden death**	Jul 1, 2022
	https://www.legacy.com/us/obituaries/dailycamera/name/catherine-fleagle-obituary?id=35943485			
444	**Zelia Guzzo**	37	**Brain bleed days post vaccination**	Mar 16, 2021
	https://www.quotidianodigela.it/zelia-non-ce-lha-fatta-dopo-12-giorni-e-morta-la-prof/			
445	**Vincent Doffont**	37	**Sudden death after training**	Mar 19, 2022
	https://hudsoncountyview.com/harrison-firefighter-vincent-doffont-dies-suddenly-after-training-exercise/			
446	**Maggie McMahon**	37	**Stroke**	Mar 25, 2022
	https://swimswam.com/riptide-swim-coach-maggie-mcmahon-dies-unexpectedly-at-37/			
447	**Marilina Calazans**	37	**Cardiac event**	Jun 2, 2022
	https://goodsciencing.com/covid/athletes-suffer-cardiac-arrest-die-after-covid-shot/			
448	**Paplu Dixit**	37	**Sudden death during workout**	Jun 25, 2022
	https://allsarkarinaukari.in/national/jharkhand-news-while-lifting-weight-in-the-gym-the-young-man-suddenly-fell-unconscious-lost-his-life-immediately-jharkhand-news-while-lifting-weight-in-the-gym-the-young-man-suddenly-fell-uncon/			
449	**Rob Wardell**	37	**Died unexpectedly in his sleep**	Aug 23, 2021
	https://www.theguardian.com/sport/2022/aug/24/mountain-bike-rider-rab-wardell-dies-aged-37			

450	**Franck Berrier**	37	**Cardiac event during game**	Aug 13, 2021

https://www.thesun.co.uk/sport/football/15860761/franck-berrier-dead-37-heart-attack/

451	**Parys Haralson**	37	**Died unexpectedly in his sleep**	Sep 14, 2021

https://sports.yahoo.com/former-saints-linebacker-parys-haralson-005433655.html

452	**Filippo Morando**	37	**Sudden death during run**	Oct 16, 2021

https://www.padovaoggi.it/cronaca/medico-muore-mentre-fa-jogging-50-metri-casa-camposampiero-padova-17-ottobre-2021.html

453	**Michael Englebert**	37	**Cardiac event after game**	Oct 25, 2021

https://www.archyde.com/the-excitement-after-the-disappearance-of-mika-englebert-la-roche-en-ardenne/

454	**Major Wingate**	37	**Sudden death**	Oct 2, 2021

https://247sports.com/college/tennessee/Article/Tennessee-Vols-Basketball-Major-Wingate-dead-37-172295879/

455	**Michael Rokicki**	37	**Sudden death**	Dec 20, 2021

https://sport.tvp.pl/57555835/nie-zyje-michal-rokicki-olimpijczyk-z-io-mial-37-lat-plywanie

456	**Damien Zemmouri**	37	**Sudden death**	Dec 18, 2021

https://actu.fr/normandie/fleury-sur-orne_14271/full-contact-la-red-team-de-fleury-sur-orne-pleure-damien-zemmouri_47371486.html

457	**Didier Ceulemans**	37	**Sudden death**	Jan 6, 2022

https://www.sudinfo.be/id437575/article/2022-01-08/didier-ceulemans-joueur-de-marbaix-en-p4-est-decede-seulement-37-ans-il-avait

458	**Jessica Wilson**	37	**Vaccine-induced death, per coroner**	Sep 7, 2021

https://www.whsv.com/2021/10/06/mother-2-dies-blood-clots-after-getting-covid-19-vaccine/

459	**Cora Walker**	37	**Cardiac event**	Mar 11, 2022

https://www.ksdk.com/article/news/local/cora-faith-walker-died-heart-condition-medical-examiner/63-ffe06b7e-1413-4726-baf5-9ce450ec25ba

460	**Om Mohapatra**	37	**Cardiac event - stroke after booster**	Jan, 2022

https://twitter.com/payalm23/status/1484549132547399681?t=nXs10DCQNlhLnUYD8oQWkQ

461	**Matt King**	37	**Sudden death**	Jul 11, 2022

https://www.msn.com/en-us/news/us/co-founder-of-meow-wolf-has-died-here-are-five-things-to-know-about-matt-king/ar-AAZuVC0

462	**Nestor Flores**	37	**Cardiac event**	Jul 13, 2022

https://la100.cienradios.com/policiales/dolor-en-san-juan-dos-hombres-salieron-de-sus-partidos-de-futbol-llegaron-a-sus-casas-y-murieron-de-un-infarto/

463	**Musa Yamak**	38	**Cardiac event during match**	May 16, 2022

https://www.wionews.com/sports/watch-unbeaten-german-boxer-musa-yamak-passes-away-after-suffering-heart-attack-during-fight-480389

464	**Marco Memenga**	38	**Sudden death during game**	Aug 5, 2022

https://rheiderland.de/artikel/41684/fussballer-stirbt-bei-pokalspiel-in-filsum

465	**Cristian Caceres**	38	**Cardiac event during game**	Aug 8, 2022

https://euroweeklynews.com/2022/08/08/union-cordillera-senior-goalkeeper-dies-heart-attack-pisco-elqui-chile/

466	**Josh Ciocco**	38	**Sudden death**	Oct, 2022

https://www.hitc.com/en-gb/2022/10/06/josh-cioccos-death-at-38-saddens-hockey-world-as-tributes-paid-to-coach/

| 467 | **Vladimir Dorozhkin** | **38** | Sudden death day of vaccination | Jul 7, 2021 |

https://vk.com/wall-101793432_661299

| 468 | **Adans Alencar** | **38** | Cardiac event | Oct 17, 2021 |

https://www.world-today-news.com/former-brusque-player-dies-after-suffering-a-heart-attack-during-game/

| 469 | **Julija Portjanko** | **38** | Cardiac event | Oct 12, 2021 |

https://www.slobodenpecat.mk/en/deneska-makedonija-se-prostuva-od-poraneshnata-reprezentativka-julija-nikolikj-beshe-lavica-vo-rakometot-i-vo-zhivotot/

| 470 | **Julia Nikolic** | **38** | Cardiac event | Oct 11, 2021 |

https://www.ihf.info/media-center/news/former-north-macedonia-womens-national-team-player-julija-nikolic-passes-away

| 471 | **Tom Greenway** | **38** | Sudden death | Nov 8, 2021 |

https://www.thesun.co.uk/sport/horseracing/16674421/jockey-tom-greenway-dead-aged-38/

| 472 | **Renato Bastias** | **38** | Sudden death during triathlon | Jan 10, 2022 |

https://www.survivethenews.com/tragic-38-year-old-athlete-dies-after-suffering-heart-attack-during-swim-portion-at-ironman-70-3-pucon-video/

| 473 | **Divya Nair** | **38** | Stroke 14 days post vaccination | Aug 2, 2021 |

https://www.newindianexpress.com/states/kerala/2021/aug/24/inquiry-ordered-after-two-women-die-from-vaccine-related-complications-in-kerala-2349078.html

| 474 | **Jenn Gouveia** | **38** | Sudden death during run | Jul, 2021 |

https://toronto.ctvnews.ca/husband-of-toronto-mother-who-died-suddenly-on-run-says-he-lost-his-everything-1.5501767

| 475 | **Sylvia Di Pietro** | **38** | Died unexpectedly in her sleep | Apr 28, 2022 |

https://www.caffeinamagazine.it/italia/silvia-di-pietro-malore-improvviso-morta-38-anni/

| 476 | **Donna Taylor** | **38** | Sudden death during run | Jun 27, 2022 |

https://www.news.com.au/lifestyle/health/health-problems/horror-as-mum-38-devoted-to-her-four-children-collapses-and-dies-suddenly-on-school-run/news-story/46fcf04adf862bdace0b51a8c4c8c8c7

| 477 | **Dr. Arran Lamont** | **38** | Sudden death | Jun 6, 2022 |

https://www.dignitymemorial.com/obituaries/stony-plain-ab/arran-lamont-10788286

| 478 | **Hayden Goodrick** | **38** | Sudden death during boat race | Jun 4, 2022 |

https://www.newportri.com/story/news/local/2022/06/08/hayden-goodrick-dies-m-32-sailing-newport-ri-race-americas-cup-sailor/7553125001/

| 479 | **Mike Salase** | **39** | Sudden death during game | Jul 10, 2021 |

https://www.nzherald.co.nz/sport/northland-rugby-league-player-mike-salase-dies-while-playing-the-game-he-loved/EZQ6SVJQ7XZDSEQPAA6CEA564I/

| 480 | **Mike Elhard** | **39** | Sudden death during run | Sep 12, 2021 |

https://www.eplocalnews.org/2021/11/23/community-steps-up-to-run-for-mike/

| 481 | **Ivo Santos** | **39** | Cardiac event | Jan 28, 2022 |

http://www.adalbertogomesnoticias.com.br/2022/01/em-pao-de-acucaral-ex-jogador-do.html

| 482 | **Everton Brilhante** | **39** | Cardiac event | Jan 7, 2022 |

https://www.jornalpp.com.br/noticias/esporte/morre-ex-jogador-sao-carlense-everton-brilhante-aos-39-anos/

| 483 | **Jesse Barata** | **39** | Died unexpectedly in her sleep | Apr 14, 2022 |

https://www.legacy.com/us/obituaries/madison/name/jesse-barata-obituary?id=34899801

484 **John Dombrowski** **39** **Died unexpectedly in his sleep** Jun 28, 2022
https://archive.ph/72tk9

485 **Dr. Gulshan Binzade** **39** **Cardiac event** Jun 20, 2022
https://timesofindia.indiatimes.com/city/mumbai/mumbai-air-india-doctor-dr-gulshan-
binzade-39-dies-after-heart-attack/articleshow/92326033.cms

486 **Dr. Keshav Sharma** **39** **Sudden death days post vaccination** Jan 11, 2021
https://trinidadexpress.com/newsextra/sat-sharma-s-son-dies-in-ireland-after-covid-
injection/article_a4222eaa-6c98-11eb-8262-6b307158bd0a.html

487 **Akeem Omolade** **39** **Sudden death** Jun 13, 2022
https://euroweeklynews.com/2022/06/14/torino-striker-found-dead-leg-pain/

488 **Kassidi Kurill** **39** **Sudden death days post vaccination** Feb 5, 2021
https://www.thesun.co.uk/news/14326252/healthy-mom-died-organ-failure-moderna-
covid-vaccine/

489 **Andres Cuervo** **40** **Cardiac event** Oct 5, 2022
https://www.billboard.com/music/latin/andres-cuervo-dead-colombian-singer-dies-
at-40-1235154231/

490 **Emma Cohen** **40** **Sudden death** Apr 8, 2021
https://people.com/movies/olivia-newton-john-mourns-the-sudden-death-of-her-cancer-
nurse/

491 **Sidharth Shukla** **40** **Cardiac event** Sep 2, 2021
https://timesofindia.indiatimes.com/life-style/health-fitness/health-news/sidharth-
shukla-death-cause-actor-sidharth-shukla-passes-away-due-to-a-heart-attack-why-
heart-attacks-are-becoming-common-in-younger-ages-and-what-can-you-do-to-avoid-
them/photostory/85859578.cms

492 **Bernardino Hancco** **40** **Cardiac event during game** Aug 25, 2022
https://www.pachamamaradio.org/melgar-futbolista-de-nunoa-fallece-en-campo-
deportivo/

493 **Rajesh Verma** **40** **Cardiac event** Apr 24, 2022
https://www.republicworld.com/sports-news/cricket-news/former-ranji-trophy-winning-
cricketer-rajesh-verma-dies-at-40-articleshow.html

494 **Vincenzo Di Grande** **40** **Sudden death** Jan 4, 2022
https://stopcensura.online/piacenza-allenatore-pallanuoto-muore-a-soli-40-anni-
improvviso-malore-fulminante/

495 **Simone Bedodi** **40** **Died unexpectedly in her sleep** Oct 10, 2021
https://corrieredibologna.corriere.it/bologna/cronaca/21_ottobre_12/parma-prima-festa-
la-promozione-poi-malore-muore-giocatore-baseball-e06a9490-2b62-11ec-89eb-
b0ba4b8c21e4.shtml

496 **Jorge Heinle** **40** **Sudden death** Nov 11, 2021
https://www.infranken.de/lk/bad-kissingen/sport/ein-persoenlicher-nachruf-zum-tod-
von-joerg-heinle-art-5332939

497 **Michal Gurajdowski** **40** **Sudden death** Dec 28, 2021
https://ligawl.pl/index.php/2021/12/29/michal-gurajdowski-nie-zyje/

498 **Ron Frederick** **40** **Sudden death** Jan 3, 2022
https://www.post-gazette.com/sports/high-school-football/2022/02/03/Ron-
Frederick-Southmoreland-High-School-football-coach-WPIAL-died-dream-job/
stories/202202030112

499 **Unnamed Person** **40** **Sudden death during marathon** Dec 9, 2021
https://wien.orf.at/stories/3120997/#:~:text=Der%2038.,starb%20wenig%20später%20
im%20Spital.&text=Der%20Halbmarathon%2DLäufer%20befand%20sich,beim%20
Burgtheater%2C%20als%20er%20zusammenbrach.

500 **Jose Gabriel Jimenez** 40 **Cardiac event during run** Mar 29, 2022
https://plumaslibres.com.mx/2022/03/29/jose-gabriel-jimenez-corredor-que-fallecio-de-infarto-fulminante-en-carrera-era-el-dueno-de-jugos-california-de-zona-orizaba/

501 **"Mikaben" Benjamin** 41 **Sudden death on stage** Oct 15, 2022
https://www.theguardian.com/music/2022/oct/16/haitians-in-shock-after-death-of-singer-mikaben-in-paris

502 **Ángel Brioso** 41 **Cardiac event during half-time** Feb 13, 2022
https://euroweeklynews.com/2022/02/16/spanish-footballer-dies-heart-attack/

503 **Silvia Di Gabriele** 41 **Died unexpectedly in her sleep** Jun 17, 2022
https://www.emmelle.it/2022/06/17/malore-nel-sonno-muore-giovane-mamma-di-tre-bambini/

504 **Jay Collins** 41 **Died unexpectedly in his sleep** Jul 31, 2022
https://www.myheraldreview.com/a/socoactive/sports/former-apaches-player-coach-jay-collins-dies-at-41/article_bdbcccd2-140b-11ed-bbe7-d79980a6bfae.html

505 **Erik Volper** 41 **Cardiac event during run** Apr 21, 2022
https://dailyvoice.com/new-york/mamaroneck/obituaries/former-larchmont-resident-who-was-counselor-dies-suddenly-at-41/831052/

506 **Adam Bounds** 41 **Vaccine-induced death, per coroner** May 31, 2021
https://www.bristolpost.co.uk/news/bristol-news/popular-devon-dad-footballer-died-6320625

507 **Badr Laksour** 41 **Cardiac event during game** Oct 17, 2021
https://www.francebleu.fr/infos/faits-divers-justice/un-joueur-de-foot-decede-sur-le-terrain-a-avignon-1634490552

508 **Hannah Purvis** 41 **Sudden death** Apr 1, 2022
https://www.edp24.co.uk/news/obituaries/obituary-runner-hannah-purvis-dies-suddenly-aged-41-8896248

509 **Deepesh Bhan** 41 **Stroke while playing cricket** Jul 23, 2022
https://timesofindia.indiatimes.com/tv/news/hindi/bhabiji-ghar-par-hai-actor-deepesh-bhan-passes-away-his-co-actors-are-shocked-and-saddened-by-the-loss/articleshow/93067685.cms

510 **Sonia Acevedo** 41 **Sudden death days post vaccination** Jan 1, 2021
https://www.dailymail.co.uk/news/article-9111311/Portuguese-health-worker-41-dies-two-days-getting-Pfizer-covid-vaccine.html

511 **Jose Gozalbes** 42 **Sudden death** Apr 12, 2022
https://euroweeklynews.com/2022/04/13/real-murcia-jose-carlos-gozalbes-died-suddenly/

512 **Laura Henderson** 42 **Cardiac event during run** Mar 27, 2021
https://www.gofundme.com/f/jmg96-lauras-memorial

513 **Alexander Siegfried** 42 **Sudden death** Sep 26, 2021
http://www.anpfiff.info/mobile/sites/cms/artikel.aspx

514 **Miguel Torres** 42 **Sudden death** Jan 2, 2022
https://www.primerahora.com/deportes/baloncesto/notas/luto-en-el-baloncesto-tras-la-repentina-muerte-del-tecnico-miguel-monchy-torres/

515 **Cecilia Teri** 42 **Sudden death during workout** Jan 21, 2022
https://www.debate.com.mx/mundo/Mujer-muere-mientras-hacia-su-rutina-de-gimnasio-20220121-0105.html

516 **Josep Maria Pijuan** 42 **Sudden death during marathon** Jan 23, 2022
https://www.world-today-news.com/a-42-year-old-man-dies-in-the-llanera-trail-mountain-race/

517 **Eligio Greco**   **42**   **Sudden death**   Jul, 2022
https://quotidianomolise.com/scompare-a-soli-42-anni-eligio-greco-lutto-a-venafro/

518 **Kate Thornton**   **42**   **Cardiac event**   Jun 28, 2022
https://www.sthelensstar.co.uk/news/20259350.tributes-pour-st-helens-woman-died-suddenly-turkey/

519 **Wendy Hagath**   **42**   **Died unexpectedly in her sleep**   Jun, 2022
https://www.thescottishsun.co.uk/news/8970090/tributes-paid-to-scots-mum-died-in-sleep/

520 **Dr. Andy Jassal**   **42**   **Cardiac event**   Mar 20, 2021
https://www.gofundme.com/f/remembering-andy-jassal

521 **Janus Januszke**   **43**   **Cardiac event during run**   May 20, 2022
https://www.foxsports.com.au/afl/respected-personal-trainer-remembered-by-afl-stars-after-sudden-death/news-story/f5f62cf2a03c8fb7a34244e4a4ce7522

522 **Mzameleni Mthembu**   **43**   **Cardiac event during marathon**   Aug 25, 2022
https://runningmagazine.ca/sections/runs-races/runner-collapses-and-dies-at-comrades-marathon/

523 **Fatih Mumcu**   **43**   **Cardiac event during game**   Apr 17, 2022
https://www.yenicaggazetesi.com.tr/gecirdigi-kalp-krizi-sonrasi-kurtarilamadi-futbol-sahasinda-kahreden-olum-532491h.htm

524 **Andrew Parker**   **43**   **Died unexpectedly in his sleep**   Nov 30, 2021
https://www.thecomet.net/news/tribute-to-football-referee-stevenage-man-andrew-parker-8563584

525 **Phil Petty**   **43**   **Sudden death**   Jul 21, 2022
https://www.greenvilleonline.com/story/sports/college/usc/2022/07/21/phil-petty-dies-south-carolina-football-quarterback/10097081002/

526 **Romina De Angelis**   **43**   **Sudden death during padel game**   Feb 12, 2021
https://www.leggo.it/AMP/italia/romina_de_angelis_morta_padel_malore_improvviso-6358981.html

527 **Clare Lipscombe**   **43**   **Cardiac event**   Jul 9, 2021
https://www.stroudnewsandjournal.co.uk/news/19470841.pga-golfer-clare-lipscombe-dies-aged-43-european-tour/

528 **Marcelo De Leon**   **43**   **Cardiac event during race**   Jan 4, 2022
https://www.fmgente.com.uy/noticias/argentino-federico-bruno-ganó-fernando-52229.html

529 **Valentin Gherebe**   **43**   **Sudden death**   Jan 31, 2022
https://mortsapresvaccination.wordpress.com/2022/02/03/la-mere-accuse-le-vaccin-davoir-tue-son-fils/

530 **Dr. James Dargin**   **43**   **Sudden death**   Oct 31, 2021
https://www.legacy.com/us/obituaries/bostonglobe/name/james-dargin-obituary?id=31241101

531 **Dr. Bret Stetka**   **43**   **Sudden death**   Aug 6, 2022
https://www.medscape.com/viewarticle/979438

532 **Gregorio Pagliucoli**   **44**   **Died unexpectedly in his sleep**   May 7, 2022
https://www.lanazione.it/arezzo/cronaca/muore-nel-sonno-gregorio-pagliucoli-44-anni-allenatore-delle-giovanili-del-montevarchi-1.7648424

533 **Giuseppe Fortunato**   **44**   **Died unexpectedly in his sleep**   Jul 31, 2022
https://calabria7.it/vibo-sotto-shock-per-limprovvisa-morte-del-suo-giovane-maestro-orologiaio/

534	**Maqsood Anwar**	**44**	**Cardiac event**	Jul 19, 2021
	https://www.bbc.com/news/uk-wales-57880399			
535	**Robert Marcys**	**44**	**Sudden death during training**	Nov 2, 2021
	https://nwk24.pl/2021/11/07/nie-zyje-robert-marcys-judoka-zaka-kielce-mial-44-lata/			
536	**Mathieu Léonard**	**44**	**Stroke during run**	Jan 6, 2022
	https://www.sudinfo.be/id437617/article/2022-01-08/arlon-pleure-le-depart-de-mathieu-leonard-un-homme-au-grand-coeur			
537	**Ian Reed**	**44**	**Vaccine-induced death per coroner**	Jul 16, 2021
	https://www.standard.net.au/story/7645756/vaccine-caused-tas-mans-death-coroner/			
538	**Kimberly Credit**	**44**	**Sudden death weeks post vaccination**	Apr 5, 2021
	https://healthimpactnews.com/2021/44-year-old-pastor-dead-after-moderna-covid-shot-wanted-other-pastors-and-african-americans-to-follow-her-example-and-take-the-shot/			
539	**Steve Nguyen**	**44**	**Cardiac event during workout**	Apr 1, 2021
	https://www.dailymail.co.uk/news/article-9561637/Healthy-paramedic-44-dies-heart-attack-working-gym-leaving-family-struggling.html			
540	**Davide Piglia**	**44**	**Cardiac event while cycling**	Apr 5, 2022
	https://www.lastampa.it/asti/2022/04/07/news/appassionato_di_bici_muore_mentre_prova_il_percorso_della_gara-2918481/			
541	**Jurgen Groothaerd**	**44**	**Blood clot while cycling**	Aug 8, 2022
	https://www.nieuwsblad.be/cnt/dmf20220812_93515856			
542	**Andrea Astolfi**	**45**	**Cardiac event**	Sep 1, 2021
	https://corrieredelveneto.corriere.it/treviso/cronaca/21_settembre_11/treviso-dirigente-sportivo-muore-45-anni-l-allenamento-84b27b18-130c-11ec-8a90-29d83ddfbd2a.shtml			
543	**Murtaza Lodhgar**	**45**	**Cardiac event**	Sep 17, 2021
	https://www.devdiscourse.com/article/sports-games/1734584-mizoram-u-19-head-coach-murtaza-lodhgar-dies-of-heart-attack-in-vizag			
544	**Julio Lugo**	**45**	**Cardiac event**	Nov 15, 2021
	https://www.cbssports.com/mlb/news/julio-lugo-former-mlb-shortstop-who-won-2007-world-series-with-red-sox-dies-at-45/			
545	**Umair Siddiqui**	**45**	**Cardiac event after training**	Nov 29, 2021
	https://www.thenews.com.pk/print/913702-squash-player-dies-of-heart-attack-after-practice			
546	**Lisa Shaw**	**45**	**Vaccine-induced death per coroner**	May 21, 2021
	https://www.theguardian.com/media/2021/aug/26/bbc-presenter-lisa-shaw-died-of-astrazeneca-covid-vaccine-complications-coroner-finds			
547	**Harper Caron**	**45**	**Sudden death**	Aug 7, 2021
	https://www.nbcdfw.com/news/local/uncle-julios-president-dies-after-being-found-at-downtown-dallas-hotel/2715093/			
548	**Mark Fraser**	**45**	**Cardiac event**	Oct, 2022
	https://www.therecord.com/news/municipal-election/2022/10/05/waterloo-region-district-school-board-candidate-dies-suddenly.html			
549	**Aidan Sharanovich**	**40s**	**Cardiac event**	Aug 22, 2021
	https://tiool.com/poland/aidan-sharanovich-pogo-shizikens-wicketkeeper-has-died/			
550	**Dr. B Behzadizad**	**40s**	**Died unexpectedly in his sleep**	Jul 13, 2022
	https://chuffed.org/project/in-memory-of-drbaharan-behzadi			

A SAMPLING OF SCHOLARLY AND SCIENTIFIC STUDIES ON SUDDEN CARDIAC DEATH IN ATHLETES

- Sports-related sudden cardiac death in a competitive and a noncompetitive athlete population aged 12 to 49 years: data from a nationwide study in Denmark
  - Published May 2014
  - Study reviewed all deaths for people 12-49 years old, 2007-2009
  - Of the total 881 sudden cardiac deaths, only 44 were attributed to athletes
  - **Non-athletes had 20 times more sudden cardiac death than athletes**

- Comparison of the Frequency of Sudden Cardiovascular Deaths in Young Competitive Athletes Versus Nonathletes?
  - Published February 2016
  - **Sudden cardiac deaths were 8 times more common in nonathletes than athletes**

- Incidence of sudden cardiac arrest and death in young competitive athletes: a 4-year prospective study
  - Published November 2021
  - Study looked at sudden cardiac arrest in US competitive athletes 2014-2018
  - The study found 173 fatalities over 4-years (and 158 survivors)

- Incidence and Etiology of Sudden Cardiac Arrest and Death in High School Athletes in the United States
  - Published November 2016
  - Reviewed sudden cardiac arrest and death in high school athletes from seven states 2007-2013
  - Included 36% of the total high school athlete population during those 7-years
  - Only 69 deaths were identified over 7-years

- Sudden Cardiac Arrest During Competitive Sports
  - Published November 2016
  - Study designed to review all out-of-hospital cardiac-arrests of athletes between 2009-2014
  - **74 sudden cardiac arrests occurred during sports, but only 16 during competitive sports**

- "The occurrence of sudden cardiac arrest due to structural heart disease was uncommon during participation in competitive sports"

- Incidence and causes of sudden death in U.S. college athletes
  - Published April 2014
  - Over the 10-year study period, 182 sudden deaths occurred, 64 likely cardiovascular

- Sudden deaths in young competitive athletes: analysis of 1866 athletes who died (or survived cardiac arrest) in the U.S., 1980-2006
  - Published March 2009
  - Based upon a large national registry covering 27-year period, the study identified an average of 44 incidents per year 1980–1993, and 66 incidents per year 2000– 2006, an average of about 5 per month

- Sports-related sudden cardiac death in Spain
  - Published March 2021
  - Reviewed autopsy results for 25 provinces in Spain 2010–2017
  - **Almost all deaths occurred in non-competitive recreational sports**

- Sudden cardiac death in athletes
  - Published March 2014
  - "The incidence of SCD in young athletes is in fact very low, around 1-3 per 100,000, but attracts much public attention"

- Sports-Related Sudden Cardiac Arrest in Germany
  - Published January 2021
  - Included competitive and recreational sports activities
  - Over a 6-year period, an average of 25 deaths per year, athletes and non-athletes

- Pathogeneses of sudden cardiac death in national collegiate athletic association athletes
  - Published April 2014
  - 45 cases of sudden cardiac death were identified from 2004 to 2008
  - The most common finding was a structurally normal heart or autopsy-negative sudden unexplained death

MORE SCHOLARLY STUDIES & WRITINGS
ON SUDDEN DEATHS IN ATHLETES

1. An autopsy study of sudden cardiac death in persons age 1-40
   https://pubmed.ncbi.nlm.nih.gov/25575272/

2. Sudden cardiac death in young athletes and nonathletes
   https://pubmed.ncbi.nlm.nih.gov/21716109/

3. Sudden unexpected death of cardiac origin in the 6 to 18 years population. Role of sports?
   https://pubmed.ncbi.nlm.nih.gov/16051073/

4. Risk of sudden cardiac death in young athletes: which screening strategies are appropriate?
   https://pubmed.ncbi.nlm.nih.gov/15331292/

5. Does sports activity enhance the risk of sudden death in adolescents and young adults?
   https://pubmed.ncbi.nlm.nih.gov/14662259/

6. Survival from sports-related sudden cardiac arrest: In sports facilities versus outside of sports facilities
   https://pubmed.ncbi.nlm.nih.gov/26299232/

7. Sudden Cardiac Death: Autopsy Findings in 7200 Cases Between 2001 and 2015
   https://pubmed.ncbi.nlm.nih.gov/27973393/

8. Sudden cardiac death: a nationwide cohort study among the young
   https://pubmed.ncbi.nlm.nih.gov/27910804/

9. Sudden cardiac death in athletes
   https://pubmed.ncbi.nlm.nih.gov/18634917/

10. Cardiac disease at risk in the young athlete
    https://pubmed.ncbi.nlm.nih.gov/24856863/

11. Sudden cardiac death in young athletes
    https://pubmed.ncbi.nlm.nih.gov/24007846/

12. Incidence, Cause, and Comparative Frequency of Sudden Cardiac Death in National Collegiate Athletic Association Athletes: A Decade in Review
    https://pubmed.ncbi.nlm.nih.gov/25977310/

13. Sports-related sudden cardiac death in Switzerland classified by static and dynamic components of exercise
https://pubmed.ncbi.nlm.nih.gov/26915579/

14. Sudden cardiac death in young athletes
https://pubmed.ncbi.nlm.nih.gov/17322504/

15. Data from a nationwide registry on sports-related sudden cardiac deaths in Germany
https://pubmed.ncbi.nlm.nih.gov/26130495/

16. Sports-related and non-sports-related sudden cardiac death in young adults
https://pubmed.ncbi.nlm.nih.gov/1825009/

17. Sport Related Sudden Death: The Importance of Primary and Secondary Prevention
https://pubmed.ncbi.nlm.nih.gov/36012921/

18. Sudden cardiac death in athletes
https://pubmed.ncbi.nlm.nih.gov/1450882/

19. Sudden cardiac death in young athletes: trying to find the needle in the haystack
https://pubmed.ncbi.nlm.nih.gov/17970016/

20. Sudden Unexpected Death Due to Myocarditis in Young People, Including Athletes
https://pubmed.ncbi.nlm.nih.gov/33347841/

21. Cardiac arrest and sudden death in competitive athletes with arrhythmogenic right ventricular dysplasia
https://pubmed.ncbi.nlm.nih.gov/9474700/

22. Sudden Cardiac Death in the Adolescent Athlete: History, Diagnosis, and Prevention
https://pubmed.ncbi.nlm.nih.gov/31199891/

23. Pathoanatomic Findings Associated With Duty-Related Cardiac Death in US Firefighters: A Case-Control Study
https://pubmed.ncbi.nlm.nih.gov/30371185/

24. Causes of sudden death in young and middle-aged competitive athletes
https://pubmed.ncbi.nlm.nih.gov/9276168/

25. Risk of sports-related sudden cardiac death in women
https://pubmed.ncbi.nlm.nih.gov/34894223/

26. Incidence of sudden cardiac death in National Collegiate Athletic Association athletes
https://pubmed.ncbi.nlm.nih.gov/21464047/

27. Causes of sudden cardiac death in young athletes and non-athletes: systematic review and meta-analysis: Sudden cardiac death in the young
https://pubmed.ncbi.nlm.nih.gov/34166791/

28. Nationwide burden of sudden cardiac death: A study of 54,028 deaths in Denmark
https://pubmed.ncbi.nlm.nih.gov/33965606/

29. Sudden Cardiac Death in Athletes in Italy during 2019: Internet-Based Epidemiological Research
https://pubmed.ncbi.nlm.nih.gov/33445447/

30. Sudden cardiac death in the young: a 1-year post-mortem analysis in the Republic of Ireland
https://pubmed.ncbi.nlm.nih.gov/19221830/

31. Sudden Cardiac Death in Athletes
https://pubmed.ncbi.nlm.nih.gov/34669953/

32. Assessment of premature ventricular beats in athletes
https://pubmed.ncbi.nlm.nih.gov/30430516/

33. Epidemiology of Football-Related Sudden Cardiac Death in Turkey
https://pubmed.ncbi.nlm.nih.gov/34684142/

34. Cardiac Screening of Young Athletes: a Practical Approach to Sudden Cardiac Death Prevention
https://pubmed.ncbi.nlm.nih.gov/30155696/

35. Sudden death from cardiovascular disease in young athletes: fact or fiction?
https://pubmed.ncbi.nlm.nih.gov/9429003/

36. Incidence and Causes of Sudden Cardiac Death in Athletes
https://pubmed.ncbi.nlm.nih.gov/35710267/

37. Sudden unexpected nontraumatic death in 54 young adults: a 30-year population-based study
https://pubmed.ncbi.nlm.nih.gov/7611149/

38. Cardiac arrest at rest and during sport activity: causes and prevention
https://pubmed.ncbi.nlm.nih.gov/32523432/

39. Sudden death and physical activity in athletes and nonathletes
https://pubmed.ncbi.nlm.nih.gov/8581570/

40. Demographics and Epidemiology of Sudden Deaths in Young Competitive Athletes: From the United States National Registry
https://pubmed.ncbi.nlm.nih.gov/27039955/

41. Sudden cardiac death among general population and sport related population in forensic experience
https://pubmed.ncbi.nlm.nih.gov/26344462/

42. Sudden death during sport activity: autopsy studies of 32 cases
https://pubmed.ncbi.nlm.nih.gov/22959438/

43. Causes of sudden cardiac death in young Australians
https://pubmed.ncbi.nlm.nih.gov/14748671/

44. Incidence of cardiovascular sudden deaths in Minnesota high school athletes
https://pubmed.ncbi.nlm.nih.gov/23207138/

45. Epidemiology of Sudden Death in Organized Youth Sports in the United States, 2007-2015
https://pubmed.ncbi.nlm.nih.gov/31013114/

46. Study of sudden cardiac deaths in young athletes
https://pubmed.ncbi.nlm.nih.gov/12793636/

47. Incidence and etiology of sports-related sudden cardiac death in Denmark--implications for preparticipation screening
https://pubmed.ncbi.nlm.nih.gov/20580680/

48. Sudden cardiac death in children and adolescents between 1 and 19 years of age
https://pubmed.ncbi.nlm.nih.gov/24239636/

49.   Pathology of sudden death during recreational sports in Spain
       https://pubmed.ncbi.nlm.nih.gov/23398926/

50.   Sudden death in sports among young adults in Norway
       https://pubmed.ncbi.nlm.nih.gov/20038839/

51.   Sudden death in young competitive athletes: clinicopathologic correlations in 22 cases
       https://pubmed.ncbi.nlm.nih.gov/2239978/

WHY DEATHS MOVED FROM THE ELDERLY TO YOUNGER HEALTHIER AMERICANS

The Group Life business is comprised of policies sold to mid-sized companies and large corporations for employers to give as a benefit to their employees. This is different from individual life policies and are accounted for under a different method. Group Life polices are repriced every 1-2 years typically and as result losses or profits show up much more quickly in the Profit & Loss statement of the insurance companies. Therefore, Josh Stirling and I focused on these divisions as excess deaths would likely show up in the results very quickly if our thesis was correct. If anyone reading this book have ever worked for a large company, you might have signed a death benefit form when you onboarded with Human Resources at your firm. Typically upon starting you fill out various form's…healthcare, security cards etc. If you were offered a death benefit in this process, it would typically be 1-2 times your base salary and you'd be asked to name the beneficiary on the form; if married you would name your spouse, or if single maybe your parents. The age that experienced the most dramatic increase in the SOA excess death were the millennials (ages 25-44) in the third quarter of 2021 which represents the months of July, August, and September.

When we first revealed the CDC data in March, I mentioned in interviews that the pushback explanation for this age group was that it was due to suicides (deaths of despair), drug overdoses and missed medical treatments-cancer screenings from lockdowns. I have already argued that the obvious three-month temporal rate of change increase in August, September and October can't be explained by the three above mentioned potential causes. It's statistically impossible that in a three-month period, all those events up-ticked simultaneously across the country. The Group Life survey being a subset of the CDC data with almost the same excess mortality rate that we found in March in the CDC data for millennials makes the naysayers excuses even more absurd at the time I broke this new SOA data to the independent news media. I argued in August and early September of 2022 that this is a population that is employed with a good enough high value job to receive a death benefit. The drug overdose argument slips away because employed people with insurance tend not to have Heroin-fentanyl habits, and tend to keep their jobs.

Secondly while suicides do happen, the deaths-of-despair argument would apply mostly to unemployed folks who were displaced from COVID, but not to this population as they are employed and certainly didn't all commit suicide together in a three-month time frame. Bottomline: This subset of the population tends to be healthier and happier in general, yet experienced the same adverse event into the fall of 2021. The mainstream media offers up ridiculous excuses, but don't even ask the question if it could possibly be the "vaccine."

SOA ON SEPTEMBER 12 REVEALS US GROUP LIFE POPULATION DIED MORE THAN GENERAL US POPULATION WHICH HISTORICALLY IS NOT THE CASE

The deductive reasoning I used in the preceding section that in general the group life population is healthier than the total US population was confirmed later when I found a SOA report from October 2016 and one published recently on September 12, 2022. Both reports when examined can lead to no other conclusion than that pre-COVID this population was healthier by a wide margin than the general population. The September 2022 report is simply stunning, and basically reports that this historical relationship flipped on its head starting in 2021 and substantially adds to the thesis that the vaccines were responsible for this increase in All-Cause Mortality.

The group life insured population typically has substantially lower mortality rates than the comparable working age group of the broader US population. There are a variety of factors that contribute to this phenomenon which include the requirement to be actively at work (therefore healthy enough to work) and insured groups tend to have more education and higher earnings than that of the uninsured working age general population.

The Society of Actuaries (SOA) published a study in 2016 that covered a 4-year period from 2010 – 2013 on group life insurance mortality. The study was expansive across the group life industry and included almost 97 thousand death claims.

Exhibits 1.11 and 1.12 of the SOA study show that the group life insured mortality rates for the working ages were generally 30 – 40% of the overall US population mortality rates for the 4-year study period. In other words, the group life insured typically experience one-third to two-fifths the rate of mortality of the general US population in any given year.

Exhibit 1.11: Insured Mortality vs. Population Mortality (Female)				
Gender	Central Age	Group Life qx	Population Data	Insured / Pop Ratio
Female	22	0.150	0.445	0.338
Female	27	0.147	0.553	0.266
Female	32	0.219	0.766	0.286
Female	37	0.305	1.017	0.300
Female	42	0.484	1.535	0.316
Female	47	0.733	2.486	0.295
Female	52	1.118	3.776	0.296
Female	57	1.631	5.234	0.312
Female	62	2.553	7.665	0.333
Female	67	4.318	12.124	0.356
Female	72	8.546	18.896	0.452
Female	77	20.485	31.225	0.656
Female	82	40.135	53.181	0.755
Female	87	86.976	95.072	0.915
Female	92	137.000	162.859	0.841
Female	97	197.689	257.228	0.769

Exhibit 1.12: Insured Mortality vs. Population Mortality (Male)				
Gender	Central Age	Group Life qx	Population Data	Insured / Pop Ratio
Male	22	0.642	1.310	0.490
Male	27	0.457	1.353	0.338
Male	32	0.493	1.496	0.330
Male	37	0.597	1.764	0.339
Male	42	0.842	2.414	0.349
Male	47	1.265	3.922	0.322
Male	52	1.990	6.122	0.325
Male	57	3.045	9.074	0.336
Male	62	4.716	12.489	0.378
Male	67	6.460	18.330	0.352
Male	72	15.581	27.794	0.561
Male	77	33.606	43.720	0.769
Male	82	65.388	72.956	0.896
Male	87	120.223	123.146	0.976
Male	92	190.725	199.676	0.955
Male	97	239.223	297.713	0.804

To reiterate what has already been discussed and shown (Table 5.7), the SOA published a Group Life mortality study in August 2022 which covers the period from second quarter (April-June) 2020 through first quarter (January-March) 2022. This newer study is specifically focused on the mortality rates during the COVID-19 pandemic period for people insured by a group life insurance policy.

Again Table 5.7 of the August study shows that the 25 – 64 year age groups experienced elevated mortality for the first 5 quarters of the pandemic, but then something catastrophic hit that age group starting in Q3 2021.

Table 5.7
EXCESS MORTALITY BY DETAILED AGE BAND

Age	Q2 2020	Q3 2020	Q4 2020	Q1 2021	Q2 2021	Q3 2021	Q4 2021	Q1 2022	4/20-3/22	% COVID	% Non-COVID	% Count
0-24	116%	124%	104%	101%	119%	127%	110%	91%	111%	3.3%	8.1%	2%
25-34	127%	132%	121%	118%	131%	178%	131%	125%	133%	13.3%	19.6%	2%
35-44	123%	134%	128%	129%	133%	200%	156%	136%	142%	23.1%	19.2%	4%
45-54	123%	127%	129%	133%	119%	180%	151%	143%	138%	27.4%	10.8%	9%
55-64	117%	123%	130%	130%	114%	153%	141%	137%	131%	24.0%	6.7%	18%
65-74	117%	115%	133%	130%	108%	131%	125%	122%	122%	18.6%	3.9%	17%
75-84	114%	114%	133%	123%	106%	119%	121%	121%	119%	14.0%	4.6%	20%
85+	112%	103%	124%	111%	92%	104%	105%	103%	107%	10.3%	-3.5%	27%
All[11]	116%	115%	129%	123%	107%	134%	126%	122%	121%	17.1%	4.3%	100%

Using the table above, we can determine that the 25 – 64 age group of the group life population experienced 125% and 140% excess mortality in the last 9 months of 2020 and the full year 2021, respectively. As we know, COVID was not a respecter of persons, and it spread across the population regardless of employment, wealth, or vaccine status. We should expect, then, that the excess mortality rates by age group should be comparable across the group life population versus the general population. Remember in March Josh and I focused on general US population All-Cause Mortality CDC data for the millennial cohort ages 25-44. Specifically we focused on the trend to a new high of 84% excess mortality into the third quarter of 2021 (July-September), the massive rate of change and how impossible it would be to explain it away by a temporal and statistically impossible simultaneous increase in suicides, drug overdoses and misses medical treatments. The August SOA study which is a separate data base than the CDC essentially confirmed our analysis of the third quarter of 2021 and real-life claims were paid out. As you already know we argued the only possible explanation was the timing of the mandates and the "vaccines."

Then the SOA published a report in September 2022 which speaks to the group life mortality experience relative to the general US population mortality experience since the start of the pandemic.

The punchline is essentially that the group life population (which is much healthier than the general US population) in 2021 experienced a massive 8% higher mortality rate than

the overall US population. What changed in 2021 to cause this normal relationship to flip suddenly?

The only thing that changed was a mandated vaccination campaign where continued employment was contingent upon compliance with no exceptions.

Analysis Below:

The study shows that, indeed, the excess mortality in 2020 for the working ages were similar for the group life population versus the general population. Table 10 of that study indicates that the 15-64 age band across the general US population experienced a 123.6% excess mortality rate in the last 9 months of 2020. Given that the 15-24 age band represents a negligible % of the overall death claims, we can compare the 123.6% to the 125% in the 25-64 age band from the group life study. Close enough to call it a draw.

Table 10
U.S. POPULATION 2020 MORTALITY: MARCH 22, 2020 TO JANUARY 2, 2021

	Actual	2017-2019	7 Yr Trend	5 Yr Trend	3 Yr Trend
Total Death Actual to Expected					
All Ages	2,728,042	122.6%	121.2%	122.0%	123.2%
Age >= 15	2,705,745	122.9%	121.5%	122.3%	123.5%
Ages: 15-64	687,553	123.6%	121.6%	123.0%	124.9%
Death A to E: Excluding COVID-19					
All Ages	2,335,954	105.0%	103.8%	104.5%	105.5%
Age >= 15	2,313,804	105.1%	103.9%	104.5%	105.6%
Ages: 15-64	612,424	110.1%	108.4%	109.6%	111.2%

Now let's see what happened in 2021. Table 11 of the September 2022 SOA study indicates that the working age group of the general US population experienced an excess mortality rate of 131.7% in 2021. The 25 – 64 old ages in the group life population experienced a 140% excess mortality rate in 2021. **What happened in the insured, employed population in 2021 to cause an outsized 8% higher mortality impact relative to the same ages across the general population??**

Table 11
U.S. POPULATION 2021 MORTALITY: JANUARY 3, 2021 TO JANUARY 1, 2022

	Actual	2017-2019	7 Yr Trend	5 Yr Trend	3 Yr Trend
Total Death Actual to Expected					
All Ages	3,457,109	117.9%	116.5%	117.3%	118.5%
Age >= 15	3,427,368	118.1%	116.7%	117.5%	118.6%
Ages: 15-64	940,768	131.7%	129.6%	131.0%	133.0%
Death A to E: Excluding COVID-19					
All Ages	2,999,332	102.3%	101.1%	101.8%	102.8%
Age >= 15	2,970,010	102.3%	101.1%	101.8%	102.8%
Ages: 15-64	796,438	111.5%	109.7%	110.9%	112.6%

In Table 12 of that same study, the author concludes something similar by adjusting the US population mortality to exclude a portion of the population over age 55 (in order to compare to the overall group life results). The conclusion was the same: the working age group had higher excess mortality in the group life space than the general US population in 2021 but not in 2020. Why??

Table 12

AGE-ADJUSTED U.S. POPULATION RESULTS

	Actual to Expected March 22, 2020 to January 2, 2021				
	Actual	**2017-2019**	**7 Yr Trend**	**5 Yr Trend**	**3 Yr Trend**
Total	1,819,856	122.5%	121.1%	122.0%	123.3%
Excluding COVID	1,570,333	105.7%	104.5%	105.3%	106.4%
	Actual to Expected January 3, 2021 to January 1, 2022				
Total	2,338,139	120.0%	118.6%	119.5%	120.7%
Excluding COVID	2,021,225	103.7%	102.5%	103.3%	104.4%

With this adjustment, the Group Life market was still proportionally worse in 2021 (22.0% versus 20.0%).

Remember a thesis in stock picking is formulated in a moment of time…then as time marches on the thesis is either strengthened or weakened by subsequent data when it emerges. The proof is in the pudding so to speak mostly appearing in the quarterly results reported by the company. The same thing holds true for my vaccine as the culprit for the increase in Sudden deaths and All-Cause Mortality in 2021 and 2022. First our CDC data that we sourced showing a spike in millennial excess deaths into the third quarter, then a SOA report that came out in August with table 5.7 confirming our CDC data, and now this SOA September 12 report comes out as I was writing this section of the book and had to add the analysis immediately. Since the CDC data presented by me on independent media in March it has been my contention that the vaccine mandates announced in late summer and the anticipation of the executive order in September of 2021 caused a massive uptake of the vaccine hesitant resulting in excess mortality among young healthy millennials ages 25-44. The vaccine to me was the only plausible explanation which we unfortunately are not allowed to address on the mainstream media. Now having set the frame that a 2016 industry study indicates that the group life insured population historically has a much lower mortality rate than the US population pre COVID (30-40%) we now are presented with the September 2022 SOA report which indicates that this same employed population flipped to an 8% higher rate of excess mortality versus its baseline in 2021 than the general US population excess mortality versus its baseline in 2021. Did the virus mutate and decide to kill much healthier people in 2021? Did the virus decide to disproportionally target the employed rather than the unemployed? Did the virus decide to disproportionally target insured employees versus uninsured employees? Obviously this evidence is very compelling that something new and novel was happening to insured, working employees in 2021 that did not happen in the years prior to 2020 when COVID was raging. Using rudimentary deductive reasoning there is only one thing changed in 2021 and it wasn't the virus which was becoming less virulent. **The employed insured population was forced to take an experimental vaccine product to maintain their employment – even if they were hesitant, or had a medical or religious objection – while those who were unemployed or retired had a choice.**

If my thesis were a stock it would be up 15%-20% on this news as September 14th, 2022, and trending higher because an increased number of people are becoming aware of this trend. They might not know the reason, but the trend is becoming undeniable at this point.

The damning part of the report (page 21) is here for your reference:

> *"Comparing these numbers to the Group Life numbers, we see that, in the last nine months of 2020, the proportional impact for the Group Life market was somewhat below the impact for the U.S. population (19.8% versus 22.6%), whereas the Group Life market had a similar impact for non-COVID deaths. (5.1% versus 5.0%). For 2021, the Group Life experience was proportionally worse (22.0% versus 17.9% and 3.6% versus 2.3%). One of the features of 2021 was that excess mortality was worse for the working ages. We see that the impact on the U.S. population between 15 and 64 was 31.7% worse, much higher than for the Group Life market. The two populations are different, with Group Life having fewer, but not zero, older people exposed to coverage. In the last version of this report, we considered age-adjusted results by using only 55% of the lives exposed above age 64. Continuing to use this adjustment, we see the following results: With this adjustment, the Group Life market was still proportionally worse in 2021 (22.0% versus 20.0%)."*

APPENDIX THREE

PHINANCE TECHNOLOGIES

Phinance Technologies is a global macro alternative investment firm. The flagship product is a Futures fund using proprietary fundamental economic algorithms to generate investment ideas with rigorous risk overlays. The fund is a conservatively leveraged fund that provides a highly liquid alternative to traditional asset classes. It has a low correlation with the equity markets and is designed to be suitable as an alternative investment, particularly in a period of global uncertainty and volatility.

Carlos Alegria is the principal responsible for strategy development. He holds an MSc and PhD in Finance from the University of Southampton, UK. While pursuing his PhD on stock market anomalies, he consulted with a global macro hedge fund on quantitative strategies. After his PhD he joined Winton Capital, a leading $15bn CTA, where he was responsible for developing event-driven strategies in both stock and futures markets. He lectured Financial Risk Management at the University of Southampton, and published in international finance journals. Prior to his career in finance, he was a PhD Physicist by training.

Edward Dowd is responsible for marketing and client relations. He has worked on Wall Street most of his career spanning both credit markets and equity markets. Some of the firms he worked for include HSBC, Donaldson Lufkin & Jenrette, Independence Investments and most notably at Blackrock as a portfolio manager, where he managed a $14 billion Growth Equity Portfolio for 10 years. While at BlackRock, he consistently generated top quartile returns on the Lipper large cap growth category, and led/participated in marketing meetings that contributed to $12 billion in asset growth.

Yuri Nunes is a researcher in quantitative investment strategies and developing of systematic trading systems. He holds a PhD in Applied Physics, MSc in Financial Mathematics, PAPCC (MSc equivalent) in Physics Engineering, and a Graduate Degree in Physics Engineering from Nova University, Lisbon. Working in both academia and industry, he is Professor with tenure at the Nova Science and Technology School.

In addition to its other pro bono work, Phinance funds The Humanity Project.

THE HUMANITY PROJECT

Recognizing as we all do that financial and political capture has derailed the original intent of our public health institutions, The Humanity Project uses independent research and rigorous data analysis to examine decision-making by regulatory institutions.

SAMPLING OF 100 PUBLISHED PAPERS ON COVID VACCINE-INDUCED CARDIAC INJURIES TO YOUNG PEOPLE

1. "The alarming onset of some cases of myocarditis and pericarditis following the administration of Pfizer–BioNTech and Moderna COVID-19 mRNA-based vaccines in adolescent males has recently been highlighted." https://www.mdpi.com/2036-7503/13/3/61

2. Be alert to the risk of adverse cardiovascular events after COVID-19 vaccination https://www.xiahepublishing.com/m/2472-0712/ERHM-2021-00033

3. Myocarditis after immunization with COVID-19 mRNA vaccines in members of the US military. "23 male patients, including 22 previously healthy military members, myocarditis was identified within 4 days after receipt of the vaccine" https://jamanetwork.com/journals/jamacardiology/fullarticle/2781601

4. Patel, Y. R. (2021). Cardiovascular magnetic resonance findings in young adult patients with acute myocarditis following mRNA COVID-19 vaccination: a case series. https://www.ncbi.nlm.nih.gov/pubmed/34496880

5. Cardiovascular magnetic resonance findings in young adult patients with acute myocarditis after COVID-19 mRNA vaccination: a case series: https://pubmed.ncbi.nlm.nih.gov/34496880/

6. Myocarditis after immunization with COVID-19 mRNA vaccines in members of the US military. "23 male patients, including 22 previously healthy military members, myocarditis was identified within 4 days after receipt of the vaccine": https://jamanetwork.com/journals/jamacardiology/fullarticle/2781601

7. "The vaccine was associated with an excess risk of myocarditis" https://www.nejm.org/doi/full/10.1056/NEJMoa2110475

8. Myo/pericarditis in a previously healthy adolescent male after COVID-19 vaccination: Case report: https://pubmed.ncbi.nlm.nih.gov/34133825/

9. McLean, K., & Johnson, T. J. (2021). Myopericarditis in a previously healthy adolescent male following COVID-19 vaccination: A case report. https://www.ncbi.nlm.nih.gov/pubmed/34133825

10. Peri/myocarditis after the first dose of mRNA-1273 SARS-CoV-2 (Modern) mRNA-1273 vaccine in a young healthy male: case report: https://bmccardiovascdisord.biomedcentral.com/articles/10.1186/s12872-021-02183

11. Hasnie, A. (2021). Perimyocarditis following first dose of the mRNA-1273 SARS-CoV-2 (Moderna) vaccine in a healthy young male. https://www.ncbi.nlm.nih.gov/pubmed/34348657

12. Young male with myocarditis after mRNA-1273 coronavirus disease-2019 (COVID-19) mRNA vaccination: https://pubmed.ncbi.nlm.nih.gov/34744118/

13. Facetti, S., Giraldi, (2021). Acute myocarditis in a young adult two days after Pfizer vaccination. https://www.ncbi.nlm.nih.gov/pubmed/34709227

14. Truong, D. T. (2021). Clinically Suspected Myocarditis Temporally Related to COVID-19 Vaccination in Adolescents and Young Adults. https://www.ncbi.nlm.nih.gov/pubmed/34865500

15. Acute myocarditis in a young adult two days after vaccination with Pfizer: https://pubmed.ncbi.nlm.nih.gov/34709227/

16. Cardiovascular magnetic resonance imaging findings in young adult patients with acute myocarditis after COVID-19 mRNA vaccination: a case series: https://jcmr-online.biomedcentral.com/articles/10.1186/s12968-021-00795-4

17. Multimodality imaging and histopathology in a young man presenting with fulminant lymphocytic myocarditis and cardiogenic shock after vaccination with mRNA-1273: https://pubmed.ncbi.nlm.nih.gov/34848416/

18. Myocarditis after immunization with COVID-19 mRNA vaccines in members of the U.S. military: https://jamanetwork.com/journals/jamacardiology/fullarticle/2781601%5C

19. Acute myocardial infarction within 24 hours after COVID-19 vaccination: https://pubmed.ncbi.nlm.nih.gov/34364657/

20. Severe and refractory immune thrombocytopenia occurring after SARS-CoV-2 vaccination: https://pubmed.ncbi.nlm.nih.gov/33854395/

21. Report of a case of myopericarditis after vaccination with BNT162b2 COVID-19 mRNA in a young Korean male: https://pubmed.ncbi.nlm.nih.gov/34636504/

22. Clinical Guidance for Young People with Myocarditis and Pericarditis after Vaccination with COVID-19 mRNA: "There is a temporal association between receiving mRNA COVID-19 vaccination and myocarditis and pericarditis among youth." https://www.cps.ca/en/documents/position/clinical-guidance-for-youth-with-myocarditis-and-pericarditis

23. In-depth evaluation of a case of presumed myocarditis after the second dose of COVID-19 mRNA vaccine: https://www.ahajournals.org/doi/10.1161/CIRCULATIONAHA.121.056038

24. Acute peri/myocarditis after the first dose of COVID-19 mRNA vaccine: https://pubmed.ncbi.nlm.nih.gov/34515024/

25. Snapiri, O (2021). Cardiac Injury in Adolescents Receiving the COVID-19 Vaccine. https://www.ncbi.nlm.nih.gov/pubmed/34077949

26. Schauer, J. (2021). Myopericarditis After the Pfizer Messenger Vaccine in Adolescents https://www.ncbi.nlm.nih.gov/pubmed/34228985

27. Myocarditis and pericarditis after COVID-19 vaccination: "The incidence rate was higher in adolescents and after the administration of the second dose of messenger RNA (mRNA) vaccines. Overall, mRNA vaccines were significantly associated with increased risks for myocarditis/pericarditis" https://www.mdpi.com/2075-4426/11/11/1106

28. Clinical suspicion of myocarditis temporally related to COVID-19 vaccination in adolescents and young adults: https://www.ahajournals.org/doi/abs/10.1161/CIRCULATIONAHA.121.056583?ub%20%200pubmed

29. Peri/Myocarditis after the first dose of mRNA-1273 SARS-CoV-2 (Modern) mRNA-1273 vaccine in a young healthy male: case report: https://bmccardiovascdisord.biomedcentral.com/articles/10.1186/s12872-021-02183

30. Hasnie, A. (2021). Peri/myocarditis following first dose of the mRNA-1273 SARS-CoV-2 (Moderna) vaccine in a healthy young male. https://www.ncbi.nlm.nih.gov/pubmed/34348657

31. Young male with myocarditis after mRNA-1273 coronavirus disease-2019 (COVID-19) mRNA vaccination: https://pubmed.ncbi.nlm.nih.gov/34744118/

32. Acute myocarditis after SARS-CoV-2 vaccination in a 24-year-old male: https://pubmed.ncbi.nlm.nih.gov/34334935/

33. Young Male with Myocarditis Following mRNA-1273 Vaccination Against Coronavirus Disease-2019 (COVID-19). https://www.ncbi.nlm.nih.gov/pubmed/34744118

34. Facetti, S., Giraldi, (2021). [Acute myocarditis in a young adult two days after Pfizer vaccination]. https://www.ncbi.nlm.nih.gov/pubmed/34709227

35. Patel, Y. R. (2021). Cardiovascular magnetic resonance findings in young adult patients with acute myocarditis following mRNA COVID-19 vaccination: a case series. https://www.ncbi.nlm.nih.gov/pubmed/34496880

36. Truong, D. T. (2021). Clinically Suspected Myocarditis Temporally Related to COVID-19 Vaccination in Adolescents and Young Adults. https://www.ncbi.nlm.nih.gov/pubmed/34865500

37. Acute myocarditis in a young adult two days after vaccination: https://pubmed.ncbi.nlm.nih.gov/34709227/

38. Acute myocarditis after SARS-CoV-2 vaccination in a 24-year-old man: https://www.sciencedirect.com/science/article/pii/S0870255121003243

39. Cardiovascular magnetic resonance imaging findings in young adult patients with acute myocarditis after COVID-19 mRNA vaccination: a case series: https://jcmr-online.biomedcentral.com/articles/10.1186/s12968-021-00795-4

40. Multimodality imaging and histopathology in a young man presenting with fulminant lymphocytic myocarditis and cardiogenic shock after vaccination with mRNA-1273: https://pubmed.ncbi.nlm.nih.gov/34848416/

41. Myocarditis after immunization with COVID-19 mRNA vaccines in members of the U.S. military: https://jamanetwork.com/journals/jamacardiology/fullarticle/2781601%5C

42. In-depth evaluation of a case of presumed myocarditis after the second dose of COVID-19 mRNA vaccine: https://www.ahajournals.org/doi/10.1161/CIRCULATIONAHA.121.056038

43. Acute peri/myocarditis after the first dose of COVID-19 mRNA vaccine: https://pubmed.ncbi.nlm.nih.gov/34515024/

44. Acute myocardial infarction within 24 hours after COVID-19 vaccination: https://pubmed.ncbi.nlm.nih.gov/34364657/

45. Report of a case of myopericarditis after vaccination with BNT162b2 COVID-19 mRNA in a young Korean male: https://pubmed.ncbi.nlm.nih.gov/34636504/

46. Clinical Guidance for Young People with Myocarditis and Pericarditis after Vaccination with COVID-19 mRNA: "There is a temporal association between receiving mRNA COVID-19 vaccination and myocarditis and pericarditis among youth." https://www.cps.ca/en/documents/position/clinical-guidance-for-youth-with-myocarditis-and-pericarditis

47. Petra Zimmerman, Andrew J Pollard (2021) "The relatively low risk posed by acute COVID-19 in children, and uncertainty about the relative harms from vaccination and disease mean that the balance of risk and benefit of vaccination in this age group is more complex." (Andrew Pollard is chair of UK's Joint Committee on Vaccination & Immunization https://pubmed.ncbi.nlm.nih.gov/34732388/

48. Das, B. (2021). Myocarditis and Pericarditis Following mRNA COVID-19 Vaccination in Children. https://www.ncbi.nlm.nih.gov/pubmed/34356586

49. Association of myocarditis with COVID-19 messenger RNA BNT162b2 vaccine COVID-19 in a case series of children: https://pubmed.ncbi.nlm.nih.gov/34374740/

50. Epidemiology and clinical features of myocarditis/pericarditis before the introduction of COVID-19 mRNA vaccine in Korean children: a multicenter study: https://pubmed.ncbi.nlm.nih.gov/34402230/

51. Epidemiology and clinical features of myocarditis/pericarditis before the introduction of COVID-19 mRNA vaccine in Korean children: a multicenter study https://search.bvsalud.org/global-literature-on-novel-coronavirus-2019-ncov/resource/en/covidwho-1360706

52. Association of myocarditis with COVID-19 messenger RNA BNT162b2 vaccine in a case series of children: https://jamanetwork.com/journals/jamacardiology/fullarticle/2783052

53. Association of myocarditis with the BNT162b2 messenger RNA COVID-19 vaccine in a case series of children: https://pubmed.ncbi.nlm.nih.gov/34374740/

54. Myocarditis associated with SARS-CoV-2 mRNA vaccination in children aged 12 to 17 years: stratified analysis of a national database: https://www.medrxiv.org/content/10.1101/2021.08.30.21262866v1

55. Association of myocarditis with COVID-19 mRNA vaccine in children: https://media.jamanetwork.com/news-item/association-of-myocarditis-with-mrna-covid-19-vaccine-in-children/

56. Epidemiology and clinical features of myocarditis/pericarditis before the introduction of COVID-19 mRNA vaccine in Korean children: a multicenter study: https://pubmed.ncbi.nlm.nih.gov/34402230/

57. Park, J. (2021). Self-limited myocarditis presenting with chest pain and ST segment elevation in adolescents after vaccination with mRNA vaccine. https://www.ncbi.nlm.nih.gov/pubmed/34180390

58. Minocha, P. (2021). Recurrence of Acute Myocarditis Temporally Associated with Receipt of the mRNA Vaccine in a Male Adolescent. https://www.ncbi.nlm.nih.gov/pubmed/34166671

59. Long, S. S. (2021). Important Insights into Myopericarditis after the Pfizer mRNA COVID-19 Vaccination in Adolescents. https://www.ncbi.nlm.nih.gov/pubmed/34332972

60. Kwan, M. (2021). mRNA COVID vaccine and myocarditis in adolescents https://www.ncbi.nlm.nih.gov/pubmed/34393110

61. Jain, S. (2021). COVID-19 Vaccination-Associated Myocarditis in Adolescents. https://www.ncbi.nlm.nih.gov/pubmed/34389692

62. Recurrence of acute myocarditis temporally associated with receipt of coronavirus mRNA vaccine in a male adolescent: https://www.sciencedirect.com/science/article/pii/S002234762100617X

63. Umei, T. (2021). Recurrence of myopericarditis following mRNA COVID-19 vaccination in a male adolescent. https://www.ncbi.nlm.nih.gov/pubmed/34904134

64. Foltran, D. (2021). Myocarditis and Pericarditis in Adolescents after First and Second doses of mRNA Vaccines. https://www.ncbi.nlm.nih.gov/pubmed/34849667

65. Chua, G. (2021). Acute Myocarditis/Pericarditis in Hong Kong Adolescents Following Comirnaty Vaccination. https://www.ncbi.nlm.nih.gov/pubmed/34849657

66. Chelala, L. (2021). Myocarditis After COVID-19 mRNA Vaccination in Adolescents. https://www.ncbi.nlm.nih.gov/pubmed/34704459

67. Azir, M. (2021). Focal Myocarditis in an Adolescent Patient After mRNA COVID-19 Vaccine. https://www.ncbi.nlm.nih.gov/pubmed/34756746

68. Guillain-Barre syndrome after COVID-19 vaccination in an adolescent: https://www.pedneur.com/article/S0887-8994(21)00221-6/fulltext

69. Myocarditis and pericarditis in adolescents after the first and second doses of COVID-19 mRNA vaccines: https://pubmed.ncbi.nlm.nih.gov/34849667/

70. Peri/myocarditis in adolescents after Pfizer-BioNTech COVID-19 vaccine: https://pubmed.ncbi.nlm.nih.gov/34319393/

71. Acute myopericarditis after COVID-19 vaccine in adolescents: https://pubmed.ncbi.nlm.nih.gov/34589238/

72. Calcaterra, G. (2021). COVID 19 Vaccine for Adolescents. Concern about Myocarditis and Pericarditis. https://www.ncbi.nlm.nih.gov/pubmed/34564344

73. Chai, Q. (2022). Multisystem inflammatory syndrome in a male adolescent after his second Pfizer-BioNTech COVID-19 vaccine. https://www.ncbi.nlm.nih.gov/pubmed/34617315

74. Myopericarditis after Pfizer mRNA COVID-19 vaccination in adolescents: https://www.sciencedirect.com/science/article/pii/S002234762100665X

75. Myo/pericarditis after vaccination with COVID-19 mRNA in adolescents 12 to 18 years of age: https://www.sciencedirect.com/science/article/pii/S0022347621007368

76. Association of myocarditis with COVID-19 messenger RNA vaccine BNT162b2 in a case series of children: https://jamanetwork.com/journals/jamacardiology/fullarticle/2783052

77. Important information on myopericarditis after vaccination with Pfizer COVID-19 mRNA in adolescents: https://www.sciencedirect.com/science/article/pii/S0022347621007496

78. Self-limited myocarditis presenting with chest pain and ST-segment elevation in adolescents after vaccination with the BNT162b2 mRNA vaccine: https://pubmed.ncbi.nlm.nih.gov/34180390/

79. Symptomatic Acute Myocarditis in 7 Adolescents after Pfizer-BioNTech COVID-19 Vaccination: https://pediatrics.aappublications.org/content/148/3/e2021052478

80. Cardiac injury in adolescents receiving mRNA COVID-19 vaccine: https://journals.lww.com/pidj/Abstract/9000/Transient_Cardiac_Injury_in_Adolescents_Receiving.95800.aspx

81. Peri/Myocarditis in adolescents after Pfizer COVID-19 vaccine: https://academic.oup.com/jpids/advance-article/doi/10.1093/jpids/piab060/6329543

82. Acute myocarditis after SARS-CoV-2 vaccination in a 24-year-old male: https://pubmed.ncbi.nlm.nih.gov/34334935/

83. Acute symptomatic myocarditis in seven adolescents after Pfizer-BioNTech COVID-19 vaccination: https://pediatrics.aappublications.org/content/early/2021/06/04/peds.2021-052478

84. Nygaard, U. (2022). Population-based Incidence of Myopericarditis After COVID-19 Vaccination in Danish Adolescents. https://www.ncbi.nlm.nih.gov/pubmed/34889875

85. COVID-19 vaccine-induced myocarditis: a case report with review of the literature: https://www.sciencedirect.com/science/article/pii/S1871402121002253

86. STEMI mimicry: focal myocarditis in an adolescent patient after COVID-19 mRNA vaccination: https://pubmed.ncbi.nlm.nih.gov/34756746/

87. Kohli, U. (2021). mRNA Coronavirus-19 Vaccine-Associated Myopericarditis in Adolescents: A Survey Study. https://www.ncbi.nlm.nih.gov/pubmed/34952008

88. Epidemiology of acute myocarditis/pericarditis in Hong Kong adolescents after co-vaccination: https://academic.oup.com/cid/advance-article-abstract/doi/10.1093/cid/ciab989/644 5179.

89. Epidemiology of acute myocarditis/pericarditis in Hong Kong adolescents after co-vaccination: https://academic.oup.com/cid/advance-article-abstract/doi/10.1093/cid/ciab989/6445179

90. Clinical suspicion of myocarditis temporally related to COVID-19 vaccination in adolescents and young adults: "Abnormal findings on cMRI were frequent. Future studies should evaluate risk factors, mechanisms, and long-term outcomes." https://pubmed.ncbi.nlm.nih.gov/34865500/

91. Myocarditis associated with COVID-19 vaccination in adolescents: https://publications.aap.org/pediatrics/article/148/5/e2021053427/181357

92. Myocarditis findings on cardiac magnetic resonance imaging after vaccination with COVID-19 mRNA in adolescents:. https://pubmed.ncbi.nlm.nih.gov/34704459/

93. Acute myopericarditis after COVID-19 vaccination in adolescents:. https://pubmed.ncbi.nlm.nih.gov/34589238/.

94. "7 cases of acute myocarditis or myopericarditis in healthy male adolescents who presented with chest pain within 4 days after the second dose." https://pubmed.ncbi.nlm.nih.gov/34088762/.

95. Myocarditis and pericarditis in adolescents after first and second doses of COVID-19 mRNA vaccines:. https://academic.oup.com/ehjqcco/advance-article/doi/10.1093/ehjqcco/qcab090/64 42104

96. COVID-19 Vaccination Associated with Myocarditis in Adolescents: https://pediatrics.aappublications.org/content/pediatrics/early/2021/08/12/peds.2021-053427.full.pdf

97. Kostoff, R. N. (2021a). "Why are we vaccinating children against COVID-19?" https://www.ncbi.nlm.nih.gov/pubmed/34642628

98. Acute myocarditis after SARS-CoV-2 vaccination in a 24-year-old man: https://www.sciencedirect.com/science/article/pii/S0870255121003243

99. Cardiovascular magnetic resonance findings in young adult patients with acute myocarditis after COVID-19 mRNA vaccination: a case series: https://pubmed.ncbi.nlm.nih.gov/34496880/

100. Myopericarditis in a previously healthy adolescent male after COVID-19 vaccination: Case report: https://pubmed.ncbi.nlm.nih.gov/34133825/

For links to 1000 (one thousand) scientific papers on COVID vaccine injury...

APPENDIX FIVE

The Malone Doctrine

A Declaration Of Independence
From The Decisions Of Institutions That Lack Integrity

We The Undersigned:

Demand that all underlying data that contributes to a body of work under consideration must be made available and must remain accessible for analysis.

Proclaim the value of knowledge to society is not determined by any given creator of information. Instead, that it is the beneficiaries of knowledge who assign value to a proposition only through thorough critique, and relentless scrutiny.

Establish the free and open exchange of information and establish as a duty the authority to serve as the custodians of all data forming the basis of our decisions.

Require the full disclosure of all sources of funding regarding any citations noted or references made pertaining to any matter under consideration.

Commit to impartiality in consideration of all analytical information and data brought before us and expect the same from all others.

Foster rigorous open debate and scrutiny in consideration of and for any matter of concern.

Shall promptly make the discovery of intellectual dishonesty or professional irresponsibility known to all.

Ensure the health, welfare and safety of any whistleblower, bringing forth and/or making public an abrogation of the beliefs held herein.

Stand in opposition to censorship and will not accept representations of parties holding within themselves values that conflict with principles of free expression.

Deny no person the right to challenge, debate, petition, redress, examine, or protest, with facts and evidence any decision of this body.

_____   _____
Dr. Robert W. Malone   Dr. Jill Glasspool Malone

ACKNOWLEDGMENTS

All books are a team effort to some degree, this one more than most.

My thanks first to Gavin de Becker, for suggesting this project to me and for his contributions to the book.

Thanks to Geoff Towle for designing and creating the visual and graphical elements of this book, and to Dylan Harnett for his key role in assembling and confirming hundreds of sample cases for the Compendium.

A special thanks to Tony Lyons of Skyhorse Publishing for his encouragement and support throughout this project. With each new book, Tony Lyons again proves his belief in freedom of speech and resistance to censorship.

Data is the key to this book's conclusions, and my team of Josh Stirling, Carlos Alegria and Yuri Nunes did superb work in helping me collect and analyze the public (and non-public) information.

Speaking of non-public, my gratitude to the anonymous individuals in the insurance industry that have pointed me in the right direction and shared their inside view of what's going on. My thanks to Drs. Robert Malone and Jill Malone for their important encouragement during my early public revelations, and for showing the public what it looks like when scientists plainly speak the facts.

Tom Lewis, Barry O'Keefe and Andrew Aker contributed their insights to The Malone Doctrine, a writing that has the power to restore integrity in medical science. Dr. Naomi Wolf and the Daily Clout Team acted as true journalists exposing and analyzing the clinical trial data that Pfizer tried hard to keep secret. Brook Jackson showed bravery as a whistleblower, publicly revealing hidden machinations in the Pfizer clinical trials. Mike Moore of the Thomas Paine podcast gave my information some of its earliest public attention.

To former BlackRock colleagues, and current BlackRock experts for their kind support in my efforts. A very big thanks to the doctors who risked their incomes and good names to speak the truth about early treatment protocols, vaccine injuries and deaths that didn't have to happen.

Robert F. Kennedy Jr. has my gratitude for the Foreword he wrote, and my enduring respect for his many tireless efforts to expose the ways Big Pharma has corrupted too many of our government institutions.

And finally, a special thanks to my loving and supportive children, Pierce, Grace and Kate, who have been even more patient than usual while my mind was a million miles away, and to their mother, Eriko, who is a kind and capable steward of their growth.

As for my own growth, I thank and honor my father, Ed, for his many hours of loving advice, and the encouragement that kept me on the journey of discovery.

TO AN ATHLETE DYING YOUNG
By A.E. Housman

The time you won your town the race
We chaired you through the market-place;
Man and boy stood cheering by,
And home we brought you shoulder-high.

Today, the road all runners come,
Shoulder-high we bring you home,
And set you at your threshold down,
Townsman of a stiller town.

Smart lad, to slip betimes away
From fields where glory does not stay,
And early though the laurel grows
It withers quicker than the rose.

Eyes the shady night has shut
Cannot see the record cut,
And silence sounds no worse than cheers
After earth has stopped the ears.

Now you will not swell the rout
Of lads that wore their honours out,
Runners whom renown outran
And the name died before the man.

So set, before its echoes fade,
The fleet foot on the sill of shade,
And hold to the low lintel up
The still-defended challenge-cup.

And round that early-laurelled head
Will flock to gaze the strengthless dead,
And find unwithered on its curls
The garland briefer than a girl's.